MW00331180

CHASING THE GHOST BEAR

ON THE TRAIL OF AMERICA'S
LOST SUPER BEAST | MIKE STARK

University of Nebraska Press | LINCOLN

The University of Nebraska Press is part of a land-grant institution with campuses and programs on the past, present, and future homelands of the Pawnee, Ponca, Otoe-Missouria, Omaha, Dakota, Lakota, Kaw, Cheyenne, and Arapaho Peoples, as well as those of the relocated Ho-Chunk, Sac and Fox, and Iowa Peoples.

Library of Congress Cataloging-in-Publication Data
Names: Stark, Mike (Journalist), author.
Title: Chasing the ghost bear: on the trail of America's lost super beast / Mike Stark.
Description: Lincoln: University of Nebraska Press, [2022] | Includes bibliographical references.
Identifiers: LCCN 2021040673
ISBN 9781496229021 (paperback)
ISBN 9781496231949 (epub)
ISBN 9781496231956 (pdf)
Subjects: LCSH: Giant short-faced bear—America. | Mammals, Fossil. | Trace fossils—America. | BISAC: NATURE / Animals / Bears | NATURE / Fossils
Classification: LCC QE882.C15 S73 2022 | DDC 569/.78—dc23/eng/20211027
LC record available at https://lccn.loc.gov/2021040673

Designed and Set in Minion Pro by Mikala R. Kolander.

For the beasts, and all of us, lost to time.

CONTENTS

ILLUSTRATIONS

Chasing the Ghost Bear

PART 1

The Beast Is Here

1

Into the Dark

No one stumbles across Potter Creek Cave. Tucked into a mountainside high above Shasta Lake in Northern California, it's a hidden place, a hole fit into a limestone hill that's so well concealed by trees and thick bushes that you don't know you've arrived until you're practically on its front doorstep. Inside are narrow twisting passages, pits, galleries of pale stalactites, slick walls, bats, millipedes, eyeless spiders, and the kind of deep darkness that leaves you disoriented and trying to remember why anyone would ever willingly part with the daylight. It's perfect for secrets.

I'd come from Tucson the day before and woke up early to drizzle and gray outside my hotel window in Redding. No matter. This would be a good day. I drove a few miles north to the local U.S. Forest Service office, a squat building in the pines just off the highway. Potter Creek Cave, once the domain of native people and ancient wild animals, is now part of a federally owned forest. Peter Schmidt, an archaeologist with the agency, had gamely arranged to get me to the cave. He'd invited two fellow Forest Service employees to join us, and I'd pulled in Russell Shapiro, a paleontologist at California State University, Chico, who was happily about to end his term as the department's chair.

We chatted for a few minutes, signed some forms over the last swigs of morning coffee, gathered our gear, and drove north to the shore of Shasta Lake to a marina where a twenty-five-foot Forest Service boat was waiting for us. With the engine fired up, we huddled in the cabin to stay out of the rain and plowed through the green-gray water with a careful lookout for logs and other potentially damaging flotsam. The artificial lake is relatively new. It was crafted by hand and machine less than eighty years ago with the construction of the six-hundred-foot Shasta Dam that pooled up the Sacramento River and drowned the smaller Pit River, McCloud River, and several meandering tributaries. From the air, the lake looks like a giant chicken foot with a few extra toes. On the water, it's placid and dismal and feels like it has been there forever. We docked on the far shore, tied the boat to a tree, and stepped onto a steep hillside.

"Poison oak," Schmidt said affably once we started scrambling our way up the wet slope. "Lots of it."

There are no trails and no guides for getting to Potter Creek Cave, so we relied on GPS coordinates and a general directional sense from a couple in our group who had been there years before. I did my best to work around the vast clumps of poison oak and keep my footing as we pushed through sharp thickets and climbed hand-over-hand over a few outcroppings of jagged gray rocks on the slope. We were rewarded with a foggy view of the lake and surrounding hills, and the general agreement that any day picking your way up a mountain, even in the steady rain, beat a day in the office hunched over a computer.

We spent about an hour moving up and across the gradient. Schmidt was the first to reach the cave's opening. "There she is," he said over his shoulder to me. "It's a pretty grand entrance."

The mouth of the cave was wide and gaping, maybe thirty feet across. In front was an uneven curtain of California bay trees, live oaks, redbud trees, and wild grapes snaking their way toward sunlight. We gathered in the cave's funnel-shaped mouth, panting and sweating and offloading our packs. The entrance was dry with a mostly dirt floor and stone walls that somehow alternated between gray and orange. In the rear was the opening to the cave's deeper interior. After a few minutes, we put on helmets, strapped headlamps on them, and started feeding ourselves into the cave. The first task was climbing a few slippery footholds to get up an uneven wall about eight feet high. From a small landing, there was no avoiding the wafts of musty organic smells swirling up from the depths, some combination of wet dirt and bat guano. And then, on the floor of that upper landing, we approached a horizontal chute about two feet high leading to the rest of the cave. It's a chokepoint of sorts and a place where I paused to gather myself. No one goes into Potter Creek Cave without making a belly crawl through this bottleneck.

"Beyond this point," one early account said, "the explorer must depend for light on lamp or candle."[1]

Plenty had come to Potter Creek Cave before me, people and animals alike. The chambers and galleries once housed a staggering trove of bones—numbering in the thousands—representing at least fifty-two species. Animals still extant were well represented in the fragments, including rats, bats, deer, and rattlesnakes. Scattered among the debris—and bound by the blackness for thousands of years—were also the remains of more than twenty extinct species, exotic and familiar: camels, horses, giant sloths, ancient bison, dire wolves, mammoths, and mastodons. Vestiges, all, from the Pleistocene, the epoch from roughly 2.6 million years ago to 11,700 years ago famous for its massive, intermittent glaciers that draped over much of North America and for the oddball assortment of oversized, outlandish creatures that stalked the landscape.

Most of the extinct species found in the cave were already known to scientists by the time they were discovered in the late 1800s and early 1900s. One was not. Its discovery, in the form of a roughed-up skull missing its lower jaw, was the first sign of the biggest flesh-eating mammal ever to walk the continent. The largest of its kind, the males, stood taller than polar bears and the mighty brown bears of Alaska and possessed jaws powerful enough to annihilate skulls and snap bones like dry twigs. On all fours, the biggest stared a six-foot person in the face. On its hind legs, it towered ten feet or more and could reach its front paws up to fifteen feet. It sported claws like knife blades and teeth built for tearing and shredding.

Trifling with this bear could certainly come with ugly, often life-ending results. Any Ice Age person lucky enough to survive an encounter would've returned to their family with a story equal parts terror, adrenaline, and awe. No one alive today knows what its roar sounded like, but it's safe to assume it curdled bowels and triggered a flight response like no other. It's no wonder that some have speculated the mere presence of this barely believable bear in the Far North may have delayed the arrival of people to the interior of North America for hundreds, perhaps thousands, of years.

There had never been a mammal quite like this bear on the continent before the Pleistocene, and there hasn't been since. It had a scrunched face and a slender, catlike body with long legs that propelled it gracefully across the land. The fact that its bones have been found from Alaska and Canada's Yukon to California, the Midwest, the swamps of Florida, and even Mexico is a testament to a well-sharpened ability to roam, adapt, and find enough calories to power a body that sometimes approached a ton. Unknown to most people today, the giant short-faced bear, as it would come to be known, was its own kind of magnificence—at least for the fleeting moments it passed on the fast-spinning evolutionary stage of the natural world.

The North American giant short-faced bear survived for more than a million years before it vanished roughly around eleven thousand years ago, part of a wave of extinctions that robbed the world of half of its large mammals as the Pleistocene drew to a close. It's one of the most hotly debated mysteries among paleontologists and has been for decades. The theft included all manner of species that seem exotic and downright monstrous to those of us on this continent today: lions, giant sloths, giant armadillos, camels, mammoths, mastodons, and saber-toothed cats.

As we climbed the wet, rocky route to Potter Creek Cave, I thought about a line from paleo-ecologist Paul S. Martin about the Pleistocene extinctions. "Without knowing it," he said, "Americans live in a land of ghosts." For reasons that were still unknown to me, I wanted to be near this vanished and mostly forgotten gigantic bear, to sniff the same dank passages it had, to put my hands on the limestone walls it had brushed against, to peel back the curtain to a time just out of reach. Maybe even to commune with the dead.[2]

With the cave's narrow chute in front of me, I lowered my belly onto the wet dirt, pushed my pack in ahead, and pulled myself forward into the dark.

2

Skull

Credit for "discovering" Potter Creek Cave, and the very first evidence of a giant short-faced bear, in 1878, is usually given to a young man named James Richardson. He was from Massachusetts and had come west, eventually taking a job at a salmon hatchery on the McCloud River in the heart of the rugged pine forests and twisting waterways of Shasta County. In the 1880 census, he listed his age as twenty-five and his job as the postmaster at Baird, the tiny settlement that sprang up around the hatchery.

The hatchery along the McCloud was something of an experiment. Although well established back east, there had never been a fish-rearing station on the West Coast until Congress supplied five thousand dollars for one in 1872. The idea was to raise salmon eggs from the Pacific and then haul them by train to the East to help save depleting runs of Atlantic salmon. Livingston Stone, a Harvard-educated former Unitarian minister turned "fish culturist" in New England, was tasked with the job. He cut an imposing figure, with copious dundrearies and a stocky build that seemed to embrace the outdoors and adventure. He was determined to make the hatchery a success, no matter the difficulties with weather, isolation, transport, or friction with the native Wintu people.[1]

The hatchery was a triumph. Within just a few years, it had produced more than five million eggs—most of which were sent to the East, but at least twenty thousand of which were shipped to New Zealand. Houses, sheds, and even a water wheel went up along the McCloud. Journalists visited, as did naturalist John Muir. Stone also survived a harrowing train crash in the summer of 1873, when a converted fruit car carrying more than a dozen aquarium tanks full of fish and two tons of ice went off the rails and into Nebraska's Elkhorn River. "The car had to fill with water before it could sink, and that gave us about five seconds to get out," Stone said later. "We went out head first. . . . I felt the water closing up around my face. Then I thought I was lost and fully expected to drown there."[2]

Had Stone drowned that day in Nebraska, it's likely that Potter Creek Cave and the giant short-faced bear might not have entered the modern consciousness when they did. Once the hatchery was up and running, Stone began leading his men on a series of recreational explorations in the mountains around the hatchery, often on Sundays. "There being no church within 50 miles, the time on Sundays was usually taken up with excursions to neighboring points of interest," Stone said.[3]

In 1874 Stone and a few companions explored two caves in the hills above the hatchery. The first was a simple affair, about thirty feet wide and one hundred feet long, "with fresh green maiden's hair growing in large clusters downward from the roof." "The other is similar and more easily reached, but has in addition a dark narrow passageway leading through the interior of the mountain to a deep perpendicular abyss . . . from the bottom of which nothing, having once fallen in, could ever escape without wings or help from the outside," Stone said, adding that it had a "peculiar terror."[4]

He and his fellow adventurers spent the day exploring the cave with the help of rope ladders, lanterns, candles, and "horrible visions of rattlesnakes and tarantulas and bottomless pits." It was an exquisite cavern with pillars, archways, yawning empty rooms, and byzantine passages. Stone later named it Baird Caverns, after

Spencer Baird, head of the U.S. Fish Commission. Today it's known as the Lake Shasta Caverns and is a popular tourist attraction just north of Redding, not far from Potter Creek Cave.

It's unclear which of those initial excursions, if any, included James Richardson, but it seems likely that he either tagged along or heard about them around the camp. Either way, his interest was piqued. In November 1878, under the headline "Discovery of a Wonderful Cave," a newspaper detailed Richardson's find. He had visited it a few weeks prior, the story said, and then he and two others returned a few Sundays later to go deeper:

> They explored it a distance of 800 yards without finding a terminus. About 350 yards from the mouth of the cave, they discovered the skeleton of a man, supposed to be that of a white man, and that it had probably lain there for twenty years. The bones were all preserved, with only the skull broken by a falling stalactite. During their exploration, which lasted from early morning until sundown, they found about thirty chambers or rooms, some of them fifty or more feet in hight [sic], and all of them beautifully ornamented with stalactites, some of them thirty feet long, which sparkled from the dim light of a candle like millions of diamonds. In one room they found a series of these, shaped like a man's hand extended downward, which, when the hand was drawn over, like running one's fingers across the keys of a piano, produced a sound something similar. They found many springs of ice-cold, sparkling water, and excellent to drink.

The story also acknowledges that "the Indians have had a knowledge of its existence for many years."[5]

One thing is clear: Richardson didn't discover what became known as Potter Creek Cave, and it's unlikely he just stumbled upon it. The cavern was used by Native people, likely the Wintu or others before them, probably for thousands of years. An atlatl—an ingenious spear-throwing device—found in the cave long after Richardson arrived was determined to be around two thousand years old. My guess, since the historical record doesn't say, is that Richardson heard about the cave, possibly from the locals, and wanted to see it for himself on one of those Sunday excursions.[6]

Whatever the case, Richardson was in Potter Creek Cave in 1878 and had lowered himself some forty feet into a pitch-black chasm later named "the pit." There at the bottom, beneath a few inches of earth and among a scattering of fallen stalactites, he found the first skull of what would later be called a giant short-faced bear.

To the extent that anyone in 1878 knew anything about extinct short-faced bears from the Pleistocene in the United States, it was the result of a single tooth. More precisely, the crown of a bear's second-to-last molar on the lower left side that was found in the sand by an army captain named Bowman along the Ashley River about ten miles north of Charleston, South Carolina. Joseph Leidy, a paleontologist and anatomist at the University of Pennsylvania, announced the discovery at the June 6, 1854, meeting of the Academy of Natural Sciences of Philadelphia. The notes devote just a paragraph to the subject, stating the molar had come from a large bear that Leidy proposed naming *Arctodus pristinus*. *Arctos* is the Greek word for "bear," *odus* comes from the Greek word for "tooth," and *pristinus* in Latin means "primitive" or "early."[7]

The tooth itself didn't stay in the grasp of history for long. After Leidy examined it, the molar was returned to Bowman and then promptly lost. But it had survived long enough to give notice—however faint—that a new, mysterious kind of bear, unknown until the moment it was found, had once existed in North America.[8]

Scientists were already well aware that monstrous bears—some capable of reaching two thousand pounds—had once lived on the other side of the Atlantic and had long been extinct. Cave bears were scattered widely across Europe and parts of Asia, thriving during the late Pleistocene before vanishing as the next age took hold. They were huge, bulky things, larger than today's brown bears, with a massive, domed skull, a dished face, thick muzzle, and startlingly large paws and claws. Surely, they could eat what they wanted but seemed often to prefer a vegetarian diet. A red pigment drawing from a prehistoric cave in France shows one with rounded, pig-like ears, hunched shoulders, and what looks to me like a wry smile.[9]

A German pastor named Johann Friedrich Esper was the first to scientifically describe the possibility of the cave bears in 1774 with his book *Newly Discovered Zoolites of Unknown Four Footed Animals*. The publication set off a flurry of contrarian speculation, including that the copious collections of large and unusual bones found in caves in south Germany were not the remains of bears but of dragons or unicorns. Esper, however, felt fairly sure he was looking at bear bones but wondered why so many individuals had died together in the caves, speculating over whether there had been a flood or a great rising of the oceans that had driven the bears to the depths of the caves where they met their doom.

"Experience continues to teach us that the wildest of beasts gather together during natural catastrophes in order to try to save themselves," Esper wrote. "Their calls to one another and some unidentifiable instinct infallibly drive the creatures of one race or species to join the others of its kind."[10]

The truth was likely more pedestrian. The caves were ideal for hibernating and escaping the long brutal cold spells that raked across the European continent each year. "Death in winter sleep was apparently the normal end for the cave bear and would mainly befall those individuals that had failed ecologically during the summer season—from inexperience, illness or old age," said one paleontologist in 1968.[11] Other bears met a more violent fate when they were ambushed by Neanderthal hunters as they emerged from their caves in the spring—at least according to one relatively new hypothesis.[12]

The protective caves were also ideal for preserving the bones, which helped scientists eventually understand how widespread and common cave bears had been. They were so copious, they seemed to exist at supernatural levels, at least until you remember the bears were around for millions of years. "The most outstanding feature of this bear is the thousands upon thousands of its bones that have been found in caves throughout Europe, from France to Russia. These bones were trampled and moved about by thousands of denning bears over the years," said one scientist.

By one estimate, thirty thousand to fifty thousand cave bears died just in Austria's Dragon Cave.[13]

The mountains of bear bones were so abundant that one writer commented in 1880 that they were being "carried off by the wagonload." And during World War I, German soldiers in need of phosphates for explosives used the bear bones and sediment in the cave as a welcome substitute after guano supplies from South America dried up.[14]

While talk of extinct bears was commonplace in Europe in the 1870s, the only notion of similar bears in the United States remained embodied in that single molar found along the Ashley River in South Carolina. It had spurred speculation in a few select scientific corners, but capturing the public's imagination by talk of a mysterious but vanished bear in the United States would take more than a solitary tooth.

On a Monday night in early November 1879, members of the California Academy of Sciences met in San Francisco. Formed in 1853, just three years after statehood, the group of intellectual luminaries had proposed to develop "a thorough systematic survey of every portion of the State and the collection of a cabinet of her rare and rich productions." By 1874 the academy was making a splash with its own museum in the old First Congregational Church in downtown San Francisco, including a woolly mammoth exhibit. The planned speaker that November night for the museum's regular fortnightly meeting had canceled. Luckily there was a special guest in town, a man named Edward Drinker Cope.[15]

Not quite forty years old, Cope was handsome and slender, with a curling mustache and eyes that always seemed to be searching. At that moment, he was in the most prolific period of his professional life and becoming one of the nation's most famous naturalists. His frenetic and highly public race against Yale's O. C. Marsh to name and describe the many dinosaurs and other ancient creatures hiding underground had brought him out west. He'd come to California that summer for the first time to see the fossil beds in eastern Oregon that his rival, Marsh, had been sifting through for several years.

On his way north, he made plans to stop on the McCloud River. In retrospect, it's not a stretch to assume he might have visited the fish hatchery—such an intriguing scientific experiment would have drawn his interest. Once there, it's conceivable that James Richardson, the man who had found the weathered bear skull in Potter Creek Cave, would've passed it along to the famous scientist. It's also possible Cope got ahold of the skull during a visit to the University of California—Berkeley, where scientists were developing an interest in fossils from the region.[16]

Whatever the case, he brought the foot-long skull with him to the California Academy of Sciences meeting a few months later in San Francisco, explaining it had been found in a cave about 220 miles to the north in the wooded wilds of Shasta County. "The specimen was said to belong to a species hitherto unknown," said a newspaper account of the meeting. "It was as large as the grizzly bear and is peculiar in its short muzzle and bull-dog face."[17]

A month later, in the December 1879 issue of the *American Naturalist*, Cope described the specimen as "the cave bear of California," adding that it was "in a good state of preservation, and demonstrates that the cave bear of that region was a species distinct alike from the cave bear of the East"—meaning the *Arctodus pristinus* that Leidy identified in 1854 from that single tooth—"and from any of the existing species."[18]

He went on to describe details, noting its hefty size and its odd proportions. In particular, it had a muzzle that was shorter and wider than other bears. He took special notice of the teeth. Although they were large—not unlike those of other bears—the specimen lacked a diastema, a common gap between molars and canines that was often used to strip leaves off branches. That was enough for Cope to separate this bear from others in its parent group, known as Ursidae. He also assumed his specimen was related to giant bears that had been found on the pampas of Argentina a few decades earlier and speculated they had wandered north from South America and eventually found their way to Potter Creek Cave.

Not all of his speculations turned out to be correct (these weren't cave bears as understood at the time), but Cope had the most important basics right. With his brief, 285-word description, a new bear—once present but now vanished—had arrived in the scientific, and American, consciousness. "The species," he proclaimed, "may be called *Arctotherium simum*."

3

A Family of Bears

The bear species whose skull Cope held that night in San Francisco had come a long way, both over land and through time. Her ancestors—scientists later figured the Potter Creek Cave bear was likely a female, after comparing its size to others found later—traced their lineage back millions of years to an animal that today we'd scarcely recognize as a bear.

Evidence of the ancestral "dawn bear" goes back more than twenty million years to subtropical Europe, where a forest-dwelling creature about the size of a raccoon probably fed on plants and insects. Its history is slim and shadowy, divined only from fragments of teeth and jaws left behind. Still, scientists have pieced together that the dainty dawn bear likely split from the canid family (progenitors of wolves, coyotes, and our canine household pets) and began its own branch, which probably spawned the family of bears that eventually spread around the world.

Time, planetary changes, mutations, and other evolutionary forces sorted the bears like all species have been sorted over millennia. Some new bears arrived and flourished, only to vanish again when things got too tough. Others adapted to their environment by sheer size or with longer claws, a flexible diet or legs capable of bursts of speed just fast enough to catch their prey. It's hard to say how many different bear species have existed during Earth's history because the fossil record is so incomplete and ambiguous, and the family tree is tangled and complex. Have there been hundreds of kinds of bears in the history of the planet? Perhaps.

Today, though, there are just eight species of living bears—and the number of subspecies remains a source of vigorous, if eye-glazing, debate. There are bears in more than sixty-five countries and on all continents except Africa, Australia, and Antarctica (although a species called the Atlas bear survived in North Africa into the 1800s). Today's bears are broken down into three subfamilies. The Tremarctinae subfamily has just one surviving member, the Andean bear in South America. It's also known as the spectacled bear for the light patches of fur around its eyes that sometimes provides a charming, professorial flair—and it often makes me think of Paddington, who arrived in London as a stowaway from "deepest, darkest Peru." The subfamily Ailuropodinae also has just one member, the giant panda in China, which has been moved in and out of the bear family by scientists for decades but now seems firmly established. The rest of the bears are part of the Ursinae, or "true bears," subfamily. They include American black bears, brown bears (which include grizzlies), polar bears, Asiatic black bears, sun bears (the world's smallest), and sloth bears.[1]

As diverse as the bear family is—and was—the species all share some important traits: large heads, small ears, massive shoulders, broad paws, and thick, unretractable claws. They come equipped with strapping muscles and flexibility that allow them to adapt easily to their environment by climbing trees, digging for food and shelter, fighting off rivals, and opportunistically finding a meal. What other family of animals can snatch a salmon from a stream, scramble up a pine tree, sniff out the smell of a seal from miles away, stand on its hind legs in search of trouble, gather berries, raid ant nests, and turn over rocks in search of tasty moths? Bears tend to be omnivorous on a sliding scale across the species, with polar bears leaning the most carnivorous, panda bears leaning the most herbivorous, and sloth bears the most insect-loving. You can also tell a lot from their teeth (most bears have thirty-two to forty-two), including the molars that act as a sort of mortar and pestle to grind up food and the canines that come in handy in killing and shredding prey. The shape and function of the skull and jaw tell their own kind of story. The panda, for instance, has giant jaw muscles and a huge skull to help break and pulverize the hard bamboo stalks so important to its survival.

Cope didn't know it that night in San Francisco, but the bear in his hands did indeed have relatives in South America. We know now that the giant short-faced bear's only living relative today is the Andean bear. And both species shared genes with another smaller, extinct bruin. The Florida spectacled bear, sometimes called the Florida cave bear, occupied parts of North America around the same time as the short-faced bears before, like them, winking out at the end of the Pleistocene. The Florida bear's existence was unknown until parts of a skull and teeth were found near a country club golf course near Melbourne, Florida, and described in 1928.[2]

So when did the first giant short-faced bear exist and where? Cope could only speculate at the time, but today the evidence points to short-faced bears' evolving in North America millions of years ago. Other related bears that started out in the New World eventually found their way to South America after the two continents connected millions of years ago at what is Panama today. Cope wasn't that far off the mark in calling the Potter Creek Cave specimen *Arctotherium*, the impressively large and pug-faced bears discovered in Argentina, but it would be decades before the details of their lineage would start to be sorted out.

4

Bone Trove

Any flicker of interest E. D. Cope's short paragraph in the back of the *American Naturalist* may have generated quickly petered out. In the decade that followed, no other specimens were found, and no bone-finding expeditions were dispatched to Potter Creek Cave.

But the bear skull sparked Cope's interest one more time. In November 1891 he published a second paper on the bear skull from Potter Creek Cave. This time he included three pen and ink drawings—views from the top, bottom, and side—and noted that it was bigger than the largest of grizzly bears and roughly the same size as the European cave bears that had been found a century earlier. He speculated about how the bears may have reached North America, either from eastern Asia or South America, and noted that their short snouts must've made a peculiar impression.

"To judge by the skull alone," Cope said, "the Californian cave bear was the most powerful carnivorous animal which has lived on our continent."[1]

Once again, the news was mostly met with silence and indifference, but Cope's mysterious bear would soon have another chance at the spotlight. As the nineteenth century gave way to the twentieth, the newly formed department of anthropology at the University of California—Berkeley set out to try to determine when the first people arrived in California and what kind of animals were around at the time. Elsewhere caves have proved fruitful for hunters of ancient remains. For the California investigators, it made sense to start with the plentiful limestone caves up north in Shasta County and to the east in Calaveras County. It wouldn't be cheap, but, happily, some of the work was funded by Phoebe Hearst, the philanthropist wife of politician and mining mogul George Hearst and mother of William Randolph Hearst. The effort paid off.

In the summer of 1902, Mercer's Cave near Murphys yielded the bones of a giant ground sloth known as *Megalonyx*, capable of growing ten feet long and weighing more than a ton. The sloths had survived millions of years, only to go extinct in the late Pleistocene, some eleven thousand years ago. Investigators speculated that the specimen they found had fallen through an opening on the surface, through a chute and onto the cave floor. Farther away in the cave, human bones were found, though they were almost certainly newer than those of the sloth.

"It appears that in this region it has been some time the custom of the aborigines to throw the bodies of their dead into such caverns as this, and in places great numbers of skeletons have accumulated. The human bones found in this cave were in such position as to indicate that they had been thrown into the first chamber through a small opening above, while the *Megalonyx* had fallen some distance below this chamber," John C. Merriam, a University of California paleontologist, wrote of the investigation in 1906.[2]

That same summer, geology student assistant Eustace Furlong made the steep hike up to Potter Creek Cave to see what secrets it might hold about the arrival of humans on the West Coast. Once inside the mouth, he wormed through the chutes and passages that abruptly terminated at the lip of a sunken chamber, whose bottom was difficult to see. The only way down was by attaching a rope ladder to a stalagmite at the top and descending into the blackness one step at a time. It was, no doubt, a heart-pounding journey. Rope ladders can be dicey, made all the worse when you can't see where you're going as you drop roughly four stories into an abyss. A missed rung underfoot or a failed grip could be disastrous, if not fatal.[3]

Furlong found the bottom of the 42-foot pit without trouble and lit up a marvelous room. It was 106 feet long, 29 feet at its widest, and 4 feet wide at its narrowest. The ceiling in one spot vaulted more than 70 feet high. In many places, great clusters of gray, drip-formed stalactites hung from the ceiling like legs without feet. On the floor, fan-shaped swaths of debris emanated from each end of the chamber and met near the middle. Stepping into it, Furlong found it was a mix of reddish pebbly clay and cave breccia along with scattered limestone rocks and broken stalactites.

Work soon began in the middle of the chamber as four-foot squares were staked out so that each one could be carefully excavated and documented. The dig, which lasted through the summer, eventually revealed stratified layers of dirt and debris acting as a sort of time capsule for each period of deposition. In the case of the first Potter Creek Cave pit, there were sometimes several feet of clay and gravel on the surface followed by a thick band of gravel and then a layer of volcanic ash, followed finally by about three feet of clay and limestone blocks and more clay. The ash—reddish yellow bits of glassy silica debris—varied from less than an inch deep to more than a foot. It was later speculated that wind may have pushed it into the cave during the Quaternary period (which encompasses both the Pleistocene epoch and today's Holocene), possibly following eruptions from one of the nearby volcanoes such as Shasta or Lassen Peak, after which it settled into a bit of standing water in the cave.

The ash was interesting, but the real show was in the other layers, where Furlong and others were meticulously picking through the rubble and noting each item found. By the end of the summer, the teeth, jaws, and other remains of more than four thousand individual animals, large and small, had been identified. It was an astounding mishmash of current and ancient lifeforms: mollusks, bats, rodents, snakes, beaver, elk, deer, tortoises, camels, tapirs, horses, mammoths, mastodons, lynx, and sloths. Once it was all sorted back at the university, the list topped fifty species, including more than twenty that were extinct. Instantly Potter Creek Cave became the richest deposit of Pleistocene bones found in California to that point.

Perhaps most strange was what Furlong found in about five feet of loose dirt on the east wall of the pit: the remnants of a giant beast with much of its skeleton nearby, as if it had keeled over in the cave and been buried by centuries' worth of earth. The astonished young student took some of the bones back to the University of California, particularly keen to show them to John Merriam, who had spent the decade making a name for himself investigating West Coast fossils. The find "made university circles open their eyes," according to one account. Furlong was immediately sent back to Potter Creek Cave to retrieve the rest of the specimen.[4]

It didn't take long for the creature to reveal itself. The thick leg bones, massive shoulders and enormous, telltale pug-nosed skull pointed in one direction. More than twenty years after Cope described the skull of a giant extinct bear from Potter Creek Cave, a second *Arctotherium simum* had been found.

William Sinclair, another student of Merriam, was also pulled in, and they sent every scrap of the bear they could find back to the paleontology lab in Berkeley. Sinclair and Furlong speculated the bear might have lived in some part of the existing cave and that some of the other bones in the pit might have been there courtesy of such a large meat-eater.

"If these animals really lived in the cave," Sinclair wrote in a report of the initial finds and nearly a year later, "many of the angular fragments of bones already referred to may be the relics of bears' feasts."[5]

Soon Merriam himself was scouring the inside of Potter Creek Cave, and the public was learning what had been found. Page 11 of the August 13, 1902, edition of the *San Francisco Call* carried a six-paragraph story about the find. It sported the breathless if hyperbolic sub-headline: "Considered the Most Valuable Find in Recent History of Science."

"The specimen sent down from Furlong was unearthed five feet below the surface of the cave floor and is in excellent condition," the story said. "Not only is the head of the ancient monster intact, with the exception of a part of the lower jaw, but parts of the vertebral column, the limb bones and even the toes and claws are intact."

Like the one described by Cope, this one was larger than any known grizzly bears, the story added. "The leg and shoulder bones, when placed in position, show the brute to have been nearly five feet high from tip of shoulder to claws."[6]

No newspaper worth its salt would miss a chance to make a splash over news that a monster had been found in the midst of their readership area. On September 14, the *Call* produced an impossible-to-ignore full-page story on page 3 of the Sunday paper, complete with a life-size photo of the sixteen-inch *Arctotherium* skull set diagonally on the page with a photo of a much smaller "ordinary bear" skull superimposed. "Greatest Skeleton of a Primeval Bear Ever Found," the headline read, followed by "Strongest, Fiercest Animal of America."

It was a masterpiece of local newspaper design that thrust the extraordinary bear into the public consciousness. Bears were still big news in the state—attacks were still written up in newspapers, and the last California grizzly wouldn't be shot until 1922. The story made it clear that grizzlies had nothing on *Arctotherium*, noting it may have stood twelve feet high on its hind legs and weighed more than a ton. "Its broad muzzle, its huge canine teeth, its great limb bones all show what it was when it lived to terrify primeval man."

The story repeated Cope's speculation that the new bear specimen "who lies in state now in the laboratory trays at the University of California" was comparable to the large bears found in South America and that the species had likely wandered north to the continent at some point. Speculation, though, had its limits, the story noted. "As to the color of arctotherium's fur, science can make no guess at all. So the public is at liberty to imagine it whatever shade of brown or of black or of white that it chooses."

"He was a genuine pioneer, this arctotherium," the story concluded, "so enterprising a one that he puts to shame the forty-niners, who believe that they own California."[7]

The ink was barely dry on the full-page story when there was word of another bear specimen found at Potter Creek Cave. By then, Furlong had gone back to school at Berkeley, and Sinclair had taken over the excavation. The third specimen of *Arctotherium simum* included just the skull and a few bones but was impressive nonetheless. The *Call* was there again on September 17, 1902, with word of an additional "rare fossil monster." "This specimen . . . is much larger than the other specimen."[8]

A crew of eight returned to Shasta County in the summer of 1903, including Furlong, Merriam, Sinclair, and Annie M. Alexander, a naturalist, explorer, and bone-finder who not only funded the expedition but would later found the Museum of Vertebrate Zoology at Berkeley. The summer's mission was to continue excavating Potter Creek Cave and visit Samwell Cave, about a mile away on a bluff above the McCloud River. Merriam and his assistant, Furlong, had been intrigued by a local legend that three Indian women once went into the cave in search of pools of "magic water." The legend went that they found two pools to bathe in, and, while searching for a third, one of the women slipped and feel to her death deep in the heart of the cave. The story's been disputed, including by one Wintu woman alive in 1903, who said anyone who fell into the cave was likely escaping an enemy rather than searching for water. Whatever the case, Merriam and Furlong were intrigued. They drew straws over who would get to enter first, knowing it would likely be another harrowing descent on rope ladders. Furlong won and went in first. Merriam later told the story:

> As he descended, Furlong described the cavern opening to the light of his candle. "It widens as I go down from a diameter of ten feet at the top to a great chamber below. And here as the ladder begins to

hang free of the wall is a sharp projecting spine of rock that thrusts itself between the ropes and makes climbing difficult." Then after a longer wait, during which he moved carefully over the rickety lower fifty feet of odds and ends of rope and scraps of wood, the hobnails of his heavy shoes could be heard grating on the stone floor. It was a critical moment, waiting to learn whether the maiden whose tragic story had led us to this adventure was a reality or only the product of fertile imagination. Suddenly, with a voice raised by excitement, Furlong called up: "There's a mountain lion at the foot of the ladder." The swift train of thought in the ensuing moment I often recall, and the various possibilities that presented themselves. To go a little less than one hundred feet down an imperfectly constructed rope-ladder to help a man without weapons fighting a puma offered little hope of real assistance. If the lion should win, would it attempt to climb up the ladder? Before any plan could be framed that seemed to have value under the special circumstances, Furlong shouted: "It is a fossil mountain lion embedded in the cave floor." Then, almost without pause, and with intonation indicating that he could hardly believe his eyes, came the words: "And here on the floor below the opening is the skeleton of the Indian maiden."[9]

Samwell Cave and Furlong's Pit, as it was later christened, was thoroughly explored in the coming years. It yielded significant finds, including bones of extinct horses, mammoths, sloths, and a sheep-like animal. And, too, fossil evidence of a giant short-faced bear.

5

Inside

I thought a lot about Eustace Furlong on my trip to Potter Creek Cave. His father was a sea captain based in San Francisco, and his older brother, Herbert, worked a stint as an assistant to Merriam on a trip to the rich John Day fossil beds in Oregon. Herbert eventually moved on and Eustace stepped in, beginning a long, storied career in paleontology.[1]

Furlong was in his late twenties when he first went to Shasta County on those summer trips. Although our visits were separated by 117 years, I couldn't help but imagine his nerves and excitement as I wiggled my way into Potter Creek Cave. I've spent much of my adult life as a journalist where on the best days I insert myself into other lives and other worlds, looking for the rush of discovering some kind of secret truth just around the next corner. It might have been much the same for Furlong, I imagined, a happy explorer descending into the dark in the hopes of finding some kernel of the past that helps explain the present. Some days it's there and often it's not. So it goes.

As we pushed our way through the two-foot-tall chute on the floor of the cave's inner entrance, stale air greeted us from the other side, a slightly sweet and musty mix of bat guano and wet dirt.

Inside was a narrow hallway that wound through limestone walls that were cool and damp. With our headlamps spraying shafts of light ahead, the rocks lit up, and handwriting appeared. We took turns reading aloud the signatures and dates: 1911, 1938, 1955, 1976, and so on. We made a few nervous jokes about who might still be in there with us. I also thought about the odd impulse to mark the walls like this, adding your name to others who had come to this remote place before you. "I was here. I existed." At some point these signatures stop being graffiti and become a historical artifact. The remains in the pit played a similar role: left-behind bones that served as a notice to future finders that they, too, had once been in this place.

I trained my light beyond the names and swept it up and down the walls, thinking about Chauvet Cave in France, where prehistoric artists had used red pigment to create simple line drawings of three small cave bears near the old entrance. Suppose someone had done the same in Potter Creek Cave? No one's ever found an ancient drawing of a giant short-faced bear, as far as I know, and that record remained intact during my visit. We nudged forward, over another hump in the floor, and found the end of the hall. From there, through a slender opening on the right, was the pit Richardson, and later Furlong, had lowered into. We took turns peering in. I jumped a little when something fluttered through the beam of my light and disappeared again. It took a second to register that it was a bat and then to marvel at its ability to navigate these impossibly complex and pitch-dark caverns.

I craned my neck and pointed my light into the quiet nothingness of the pit. Over the years, it had been excavated down twenty-five feet in some places. The incomplete remains of several more short-faced bears were ultimately discovered down there. Most were smallish, leading to speculation that they were females. But had they really lived in the pit? It would have taken incredible effort, even for a physically skilled animal like a bear, to regularly climb in and out of this sunken chamber that required rope ladders for human visitors. The answer to their presence there was likely in the splays of rocky material on the pit floor and the chimney-like chutes above each one. Sinclair later surmised that much of what had been found on cave floor came through two openings above the pit where rainwater rushing down the mountain had spilled into the cavern, carrying in whatever outside material was in its path. The openings, now closed, may have been unblocked for tens of thousands of years, hence the vast layers of deposits that Furlong, Sinclair, and others sifted through during their excavations. It's possible that these great bears fell into Potter Creek Cave through those same holes, or that they had a den in some other part of the cave, somewhere more easily accessible, and that when they died their bones were eventually washed downhill into the deepest pit.

Scattered bones are one thing, but I remained fascinated with the first specimen Furlong had found, the one along the east wall of the chamber with much of its skeleton collected in one spot, indicating it may have died there. How had it arrived? I imagined this bear, trotting along the mountainside in pursuit of a smell on the wind, suddenly tumbling through a hole in the ground, feeling itself instantly airborne in the cave's cool and dark atmosphere before hitting the floor in shock and agony. I suppose its last moments may have been in the pitch blackness, panting and snorting and struggling to stand before finally giving in. Those death sounds would've certainly reached the lip of the pit where I was standing and looking in.

The truth is we don't know how those bears ended up in the pit of Potter Creek Cave. Many of the bones are now at the University of California—Berkeley, carefully catalogued, notated, and stored away. Someone someday may well put that story together. That's what paleontologists like Russell Shapiro do.

Before we went into the cave, Shapiro, the paleontologist from California State University—Chico, showed me a photo on his mobile phone of a tooth that someone had brought him a few weeks back. It was thick and spanned much of his hand. Shapiro speculated that it was the canine of a giant short-faced bear, but there hadn't been time yet to fully investigate. "I'm not sure what else it could be," he told me. His work tends to focus on more ancient and sometimes extraterrestrial territories—metamorphic geology and complex carbon chains hiding in the sedimentary record—but it's not wholly unconnected to our trip to Potter Creek Cave in search of extinct bears.

"Paleontology is hunting without killing," he wrote to me later.

We get to become voyeurs in an ancient world and spy on ecological reactions. We also get to play detective, unraveling the distortions and missing data to create realistic stories of the ancient world. But, at its heart, is the hunt—the thrill of finding bones or shells that are hidden in the landscape. While there is still that classic image of the cowboy hat and sweat-drenched shirt in the badlands, modern paleontologists also hunt with highly accurate geochemical tools and electron microscopes as well as immense databases. It is the ability to play all these roles that inspires me as a paleontologist.

I got the feeling Shapiro was also the kind of scientist keen for adventure and any chance to get into a place he'd never been, perhaps to learn something new, perhaps for the simple thrill of climbing into a darkened cave on a rainy day in the spring with a few other curiosity-seekers. Still, he knew about *Arctodus simus*, as the giant short-faced bear is known today, and marveled over its sheer size. I also asked him what he thought it'd be like to run into such a specimen.

"I would hope its short snout would not detect how easily I would foul my shorts!"

It was raining even harder when we reemerged into the afternoon light, blinking away the blackness to begin our descent from Potter Creek Cave. Hoods up and backpacks across our shoulders, we began picking our way back through the slick bushes, poison oak, and rocky reefs. We stopped in a smaller, funnel-shaped cave about fifteen minutes later. There were hard-to-find salamanders in there as well as delicate snails, so we were careful about where we stepped into the red-brown dirt. Those would be mysteries for another day, we decided, and kept snaking down the mountain toward the lake that was there but invisible through the gray and rain. On the boat ride back, we stopped to look toward Lake

Shasta Caverns, the cave Livingston Stone had explored. And we steered past where the fish hatchery at Baird used to be on the shore of the McCloud. It's now buried beneath several hundred feet of Lake Shasta, and a few of us fantasized about what we might see if we could go diving there someday. We talked, too, about what the landscape might've been like when the giant short-faced bears were here more than eleven thousand years ago. The temperatures might've been a little colder, the vegetation a bit sparser, but it wouldn't be unrecognizable. The mountains we studied from the cabin of the boat were the same ones the bears would've climbed, using their lanky legs to navigate across the slick rocks and their noses to guide them toward the prospect of a fresh meal. They might've stood tall next to that tree right there, hoping to get a better view of what was not so far in the distance.

It sunk in that the giant short-faced bears haven't been gone all that long, really. Eleven thousand years seems like an eternity for us—think of what has happened in the turbulent, remarkable course of human events during that time—but it's really a fraction of a millisecond of the 4.5 billion years of our planet. *These bears were just here*, I thought as the boat puttered through the lake chop. And then, poof, they were gone. The thought left me a little melancholic. *Sorry I missed you.*

I was drenched, dirty, and reeking of bat guano when I got back to Redding, with a several-hour drive ahead of me that evening. I stopped in a grocery store parking lot, stepped out of my rental car, and peeled off my wet shirt, pants, and shoes. An employee collecting shopping carts didn't seem to care that I was there in my skivvies and socks and neither did I. It wasn't until I was one leg into dry pants that I remembered something I'd spotted before we left Potter Creek Cave. A few feet inside the entrance was an unmistakable pile of fresh black bear scat chock-full of fur bits from some hapless creature that had been a good meal a day or two before. It made me smile. The cave may be owned by the U.S. Forest Service now, but it doesn't really belong to any one of us. It's still the domain of bears.

6

Into the Pleistocene

In the summer of 2014, Julie Meachen found herself dangling in a rope harness eight stories above the floor of giant bell-shaped cave in northern Wyoming. Moments before, securely tied off, she'd climbed down a rickety, decades-old ladder just inside the cave's entrance, which was essentially a gaping black hole in the ground some fifteen feet across. From that tiny landing, the only thing left to do was to lower herself into the void and trust the ropes and rigging.

"It's basically like stepping off backwards into the abyss," Meachen told me later. "I'd never done anything like that before. I was terrified but I wasn't going to let it stop me from going in."

Meachen, a Des Moines University paleontologist whose research typically focuses on Ice Age mammals, had waited years—including three rounds of funding requests along with paperwork and a mountain of logistics—to get into Natural Trap Cave. And within seconds of starting her eighty-foot descent, she traded the warmth of the summer sun for the cool, humid air of the darkened cave, passing effortlessly from one world into the next, from modern time into a time just beyond reach.

Natural Trap Cave, tucked into the Bighorn Mountains along the Wyoming-Montana border, had been closed with a locked gate for nearly thirty years, keeping thousands of prized pale-ontological finds out of reach for scientists like Meachen trying to piece together a picture of the late Pleistocene. Going back at least forty thousand years—some have speculated one hundred thousand years—the cave had collected the bones of thousands of animals who'd had the awful luck of accidentally falling into the limestone-walled chasm, never to return to the surface again. The hole seemed to suck in every manner of living thing within miles, including mammoths, ancient horses, camels, bison, sheep, American cheetahs, lions, giant short-faced bears, rabbits, pikas, and even birds. The reason it was such an indiscriminate killer was that it was situated on a ridge, right along a trail heavily used by animals and hidden just behind a small rise. "If you were running for your life, like a prey animal is, and you're just running and try-ing to get away, you're not going to see it coming," Meachen said.

Once over the edge, it was a terrifying free fall to the bottom. The most fortunate died upon impact. The less fortunate crashed in a heap and suffered broken limbs, internal damage, and organ failure and were left to limp around in the inescapable cave bottom until they succumbed to their wounds, starvation, thirst, expo-sure, or some combination.

Over the centuries, the bodies piled up, flesh and fur rotted away, and layers of sediment and pollen blew in, burying the car-nage and hiding the bones. The cave's soil, cool temperatures, and humidity served as an ideal storage refrigerator. The whole thing took on the look and feel of a death-cradling cathedral with its towering, vaulted ceilings and occasional shafts of deific sunlight blazing all the way from the surface to the floor.

Although no human remains have been found in Natural Trap Cave, it's likely it was well known by native people in the area. European settlers found their way to it too, eventually. Cavers and curious locals lowered themselves down by rope in the 1920s, and possibly earlier, Meachen said. In the summer of 1969, the National Speleological Society held its annual meeting in nearby Lovell, Wyoming, and participants from around the country and the world were invited to explore the cave.[1]

It was obvious that the cave was rife with old bones, but a true, in-depth paleontological investigation didn't begin until the early 1970s in a project headed by Larry Martin from the University of Kansas and B. Miles Gilbert from the University of Missouri. Each summer for more than a decade, crews labored meticulously in the musty cave, working in grids and pulling out as many bones as they could find and extract from the sediment. By 1978, when Martin and Gilbert published a research paper on the project, they announced the recovery of more than twenty thousand bone fragments and teeth from the cave, with a staggering variety of animals represented. A cranium and other parts of at least two giant short-faced bears were found several feet down in a layer of mottled blue and orange clay along with the bones of horses, sheep, camels, wolves, bison, American lions, and cheetahs. At the time, the researchers estimated that layer dated to roughly 23,900 calendar years ago. Above that layer, in grayish-brown sediment, were mammoths, pikas, ground squirrels, and collared lemmings. The mammoths dated back to around 16,700 calendar years ago. (A quick note about dates: Scientists often rely on radiocarbon dating to determine the age of bones, plants or other formerly living things that absorbed carbon during their life. It's an estimate based on how fast a carbon isotope, called carbon-14, decays. But there's not an exact match between one carbon year and one of our calendar years—and the science continues to evolve on how best to calibrate the two. In telling the story of *Arctodus simus*, and quoting from scientific research, I try to make clear which one applies, either as "calendar years" or "radiocarbon years." At times I have used calibration tables to convert radiocarbon years to calendar years for clarity.)[2]

In 1993 a more complete survey of the excavated remains from Natural Trap Cave was released, including that there was evidence of at least 83 individual horses who'd fallen into the cave. They, along with 47 bighorn sheep, made up about 57 percent of the species found. The researchers also reported 2 camels, 6 bison, and 29 wolves.[3]

Most of the specimens found their way to the collections at the University of Kansas. After the summer of 1985, the gate erected above the cave's entrance in the 1970s was closed and locked for good, or so it seemed.

Meachen, doing postdoctoral work, first heard about Natural Trap Cave years later at a science symposium in a conversation with a researcher at an Australian university. He studied ancient DNA and was looking to team up with a North American paleontologist to get into the cave to retrieve nuclear DNA. It took several years and three attempts to secure funding, but it finally came through. By the summer of 2014, she was at Des Moines University and now the sole leader of the expedition into the cave, helping to oversee twenty to fifty researchers, cavers, and volunteers on the project.

"It just snowballed into this all-encompassing thing," Meachen said.

Despite the terrifying first trip down the ropes to the cave bottom and the near-constant temperatures of about forty degrees, she quickly grew accustomed to it and began digging, often picking up where the earlier researchers had left off. When Meachen and I first spoke, she'd been into Natural Trap Cave over the course of four summers and was hoping to do two more trips down. Perhaps most exciting was the sheer number of bones and other specimens buried in the dirt. "It's really exhilarating," she said. "It's also very, very cold down there."

One of the things that she and others have noticed is that Natural Trap Cave tended to collect fast-running species—predators such as lions and cheetahs along with prey such as horses and sheep—rather than the bigger, more plodding animals like mastodons or giant sloths. Of particular interest to Meachen were the wolves. Until her team came along, it was unclear exactly what kinds of wolves had fallen into the cave. The choices were two extinct species, Beringian wolves (thought to be mostly from the Far North) and dire wolves (which tended to be found farther south), and the smaller kind of gray wolves that still exist. She and her coauthors compared the mandibles from Natural Trap Cave with others found elsewhere. And then came a surprise from their 2016 findings: the wolves found in the Wyoming cave weren't normal gray wolves and they weren't dire wolves. They were more like

Beringian wolves, making Natural Trap Cave the farthest south those wolves had ever been found. In a way, it made sense, since there was once a well-documented gap in the giant ice sheets from the late Pleistocene that could've funneled those wolves directly from today's Alaska to Wyoming and beyond.[4]

And, of course, the wolves had company. It's no surprise that giant short-faced bears were lumbering around the landscape, which wasn't all that different then from how it looks today: mountains surrounded by open sagebrush country. Their long legs allowed them to move fast and far, and their stubby snout and powerful jaws would've been a deadly combination for other, smaller animals. "I think they were serious opportunists just like most bears today are," Meachen said.

That meant there would have been plenty of opportunity for drama in the area around the cave. "They probably would've fought over carcasses. When one kills something, the other comes in and says, 'I'm taking this for me,'" Meachen said. "I'm sure that happened. And if you've got twelve wolves there, then the bear is going to say, 'I'm not going to have anything to do with this.' There was probably a lot of antagonistic competition."

It was nothing, though, compared to the large-scale spectacle unfolding in the world around them.

In 1876 the British naturalist Alfred Russel Wallace, a colleague of Charles Darwin and cofounder of modern evolutionary biology, published a book looking at the distribution of animals around the world and piecing together the story of wildlife extinctions, something that was still very much a work in progress among scientists. Nevertheless, he understood that the planet was a fundamentally different place than it had been just a few thousand years before. "It is clear, therefore, that we are now in an altogether exceptional period of the earth's history," Wallace wrote. "We live in a zoologically impoverished world, from which all of the hugest, and fiercest, and strangest have recently disappeared."[5]

The high drama of the Pleistocene had been a long time coming. The planet was a warmer place in the preceding epoch known as the Pliocene, a time between roughly 5.3 million years ago and 2.6 million years ago. The Arctic was ice free, and deep, green boreal forests reached as far north as northern Greenland. Continents continued to shift and the isthmus at Panama emerged from the water, allowing species to move between North America and South America, a development called the "Great American Interchange." Dinosaurs were of course long gone by then thanks to the meteor that crashed into the Yucatan Peninsula some 65 million years ago. Now mammals flourished, including long-legged grazers that feasted on the spreads of savannas and grasslands.

For reasons still debated, the Earth cooled rapidly around 3.3 million years ago, and great ice sheets clawed their way across parts of North America, Greenland, Europe, and Asia. The temperatures warmed after several hundred thousand years, only to tumble again into a deep cold thousands of years later. Thus began, around 2.6 million years ago, the topsy-turvy cycles of the Pleistocene, a period described by paleontologist Ross D. E. MacPhee as "a global-scale roller coaster—full of leaps and dives, some sharp, others gentle, and surely harrowing for the unprepared."[6]

The whiplash between warm and cold (not to mention the rise and retreat of vast walls of glacial ice) unleashed vast vegetative changes across the planet that in turn set off ripple effects among animals and plants—only to be reversed when conditions changed back many years later. "With each transition between interglacial and glacial, and vice versa, the Earth's plant communities underwent violent changes," Richard Leakey and Roger Lewin wrote. "When global temperatures plummeted, tropical forests fragmented and shrank, and forests and woodlands at high latitudes migrated toward the equator. Because animal species depend on plants for their survival, either directly or indirectly, they migrated too, if they were able. At the beginning of the interglacials, the reverse process ensued. The Pleistocene generated pulses of global biotic turmoil, in slow motion."[7]

Sometimes I imagine seeing the Pleistocene play out in a time-lapse video: gigantic grinding, blue-veined glaciers growing and contracting, plants coming and going as growing conditions swing to and fro, grasslands turning to forests and back again, animals large and small doing their best to keep up, desperately shifting their ranges as the world changes around them. Cold-adapted species like mammoths and steppe bison would've proliferated across the landscape when the cool climate flourished and retreated to lower-temperature refugia when things warmed up, making room for camels, giant sloths, giant beavers, and other warmer-adapted herbivores to take their turn expanding on the land. These were boom and bust cycles—good times and bad depending on who you were, where you lived, and what the climate was like. Hungry meat-eaters like wolves and giant bears would do their best too to ride out each wave and shift their diets as necessary. Times were always uncertain and those who adapted best survived.

Finnish paleontologist Björn Kurtén called the Pleistocene "a major crisis in the history of life. Mountains are higher than ever, land areas wider, climates colder, animals larger. Climatic convulsions shake the earth; gigantic ice sheets turn immense tracts into frozen waste. Meanwhile early man proliferates and spreads to the end of the habitable world. At the same time the crisis mounts: the Age of Mammals is coming to an end; death and destruction succeed the exuberant Pliocene proliferations of life."[8]

"The pace of evolution," he added, "which had been rather slow in the Pliocene, now markedly increased."

In South America one type of ground sloth ballooned to 10,000 pounds. Australia became home to oversized koalas, giant kangaroos, and wombat-like animals as big as rhinos. Giant sloth lemurs on Madagascar may have reached 350 pounds, and their neighbor, the giant elephant bird, probably tipped the scales at 500 pounds.

The show was just as startling in North America. Columbian mammoths reached 13 feet tall and more than 22,000 pounds, something akin to the size of a moving truck. Mastodons were a little smaller, reaching around 9 feet and weighing around 10,000 pounds. There were also giant beavers, heaving glyptodonts (relatives to modern armadillos but much larger), giant peccaries, giant sloths, oversized camels, and stilt-legged llamas. *Smilodon*, the saber-toothed cat, developed a special flesh-cutting tool embedded in its deathly smile: serrated teeth that could reach nearly a foot long. All were doing whatever they could to survive the inhospitable world they found themselves in.

During the warm periods, temperatures weren't all that different from what they are today, and it's easy to conjure up bucolic scenes full of sun, plentiful water, and vast, nourishing grasslands hosting copious and happy herds of ridiculously large animals. The last big warming period of the late Pleistocene was 123,000 to 130,000 years ago. The cold periods were more stark and brutal. Many of the grasslands evolved into boreal forests, pushing those big grazing and browsing beasts ever onward in search of grasses and other fresh food. Predators, forever tied to those prey species, were forced to move too. The margins between life and death narrowed as resources dwindled and bitter cold temperatures took a toll, especially on the young and the weak. The world became more barren, inhospitable, and difficult to navigate. During the Last Glacial Maximum (23,000 to 27,000 years ago), mind-boggling ice sheets, some towering ten thousand feet tall, covered most of northern North America, except parts of Alaska and the Yukon.[9]

Mammoths, camels, horses, bears, wolves, and others somehow found a way to make do through the perils of the Pleistocene. They adapted as the world changed around them, gritting into the wind as they searched for new homes, grew thicker coats, and adjusted their diets when needed. But around fifteen thousand years ago, something began to change. The twilight of the Pleistocene took a much darker turn for larger mammals, including those in North America. By the onset of the next epoch, known as the Holocene, many would be gone in a geological blink of an eye.

7

Bears in Proximity

The first bear I ever saw in the wild was when I was six or seven and our family was camping in Washington state, somewhere in the Cascade Mountains. We had a used pop-top camping trailer that we towed behind the station wagon, the kind where each end unfolded like a pair of canvas wings and with barely enough foam-pad mattresses to accommodate the five of us. Around dusk, just as dinner was coming together, a family of black bears wandered through the campsite. My dad, maybe feeling like the rickety trailer wasn't quite protective enough, herded us back into the car. From the safety of the old Volvo, we watched as the mama black bear and her two cubs walked through camp like they owned the place. One of them punched a hole in our plastic water container and another knocked over the Coleman cooler in pursuit of whatever was inside. I don't remember anyone in the car feeling scared, but no one was interested in chasing them away either or doing anything other than letting them have their way with our belongings. These bears seemed to have a special power over us. Within a few minutes, they slipped into the cover of the woods.

I saw more bears over the years but didn't have another close encounter until the fall of 1999 in Montana. I had taken a couple weeks off from my job at an Oregon newspaper to cruise around the West to camp, hike, visit friends, and see places I hadn't seen.

On a Sunday afternoon, I was at Missoula coffee shop and unsure where I'd stay the night. One of the young guys behind the counter tipped me off to an abandoned campground southeast of town in the Bitterroot Mountains. It was just off the interstate, not far from the site of the annual bovine culinary gathering known then as the Testicle Festival. Find the sign for that, get off the highway, and the campground is just a few miles more, he told me. I found the campground tucked on the valley floor between two mountains, heavily forested with conifers and, happily, devoid of any other human beings.

It was nearly dark by the time I finished pitching the tent and lighting the fire. After hot dogs and a cup of whiskey I grabbed my flashlight and went for a nighttime walk through the empty campground. A few minutes later I nearly walked right into a black bear rooting through an old garbage can. As I stumbled backward, the beam of my flashlight passed over it long enough to tell it wasn't a very big bear, probably a juvenile on the prowl for an evening snack. We'd startled each other and then locked into a stare for a second or two. My biological responses kicked in: short breath, the white noise of blood rushing in my ears, and my heart beating like it needed out of my chest. The bear, just as scared I'm sure, made the first move, hopping out of the garbage can with a strange low growl, and then there was a sudden clawing at the ground like it was running in place just before it gained traction and took off toward me.

Bear experts typically advise you to stand your ground against black bears, lest your flight response trigger their own impulse to give chase. Of course, I did just the opposite. Without thinking, I turned and ran down the gravel path, concentrating on making my legs move as fast as possible and waiting for the swipe of claws across my calves that would crumple me to the ground in my final, grisly seconds of life on Earth. Surely there were worse ways to go, but I couldn't think of any just then. I was too focused on my white-hot pursuit of survival, alone in the woods and far from home.

The chase was probably over a few moments after it started. I think the bear ran after me in a short burst of speed, just enough to scare me off its snack, and then stopped and watched as this clumsy bipedal, reeking of hot dogs and fear, disappeared into the trees. For my part, I gave the bear and the trash can an extremely wide berth as I tromped through the dark woods on the way back to my campsite. I sat by the fire for a few more hours, my ears now attuned to every snapping twig in the blackness around me, sure that I was being stalked and unlikely to make it to sun-up. My thoughts toggled between *I'm prey to this bear, flesh and protein, nothing more,* and, more to the actual point, *you're an idiot, you got spooked by a little bear and now you're acting like a ridiculous baby.* I spent the rest of a long cold night in the driver's seat of my Subaru, doors locked, a nervous fool waiting for first light.

Still more bears came into my life. A couple years later, I moved to Cody, Wyoming, to be the one-man bureau for Montana's *Billings Gazette*, ready to entertain freezing winters, ceaseless wind, and quiet small-town nights in exchange for Yellowstone National Park, a short drive away, as my primary beat. My soon-to-be-wife and I had spent the better part of a year driving around the country in a 1984 Volkswagen Westfalia, doing our best to avoid regular employment and scraping by on a trickle of freelance writing jobs. Cody, about an hour's drive east of Yellowstone, was as good a place as any to land when the money finally petered out.

Within a few weeks of my first byline at the *Gazette* in early 2002, I went to see a dead grizzly outside of town. He was a male, about five years old, weighing around 315 pounds. Local wildlife officials knew him as no. 380 and, lately, he'd come out of hibernation hungry and had broken into a cabin in search of food. He was habituated to human food—some locals had even been feeding him apples—and had also tried breaking into someone's pickup. "We consider breaking into a building a pretty serious deal. That's usually where we draw the line," Mark Bruscino, a bear management officer with the Wyoming Game and Fish Department, told me at the time.

Fearful the bear had become too acclimated to people and now posed a danger, the wildlife crew captured 380, tranquilized him, and then injected him with a lethal cocktail. I met Bruscino at the Game and Fish office, and he led me to the garage where the bear was laid out on the concrete floor. I knelt down and put my hands into his thick dark fur, lifted one of his meaty arms, and examined his squishy paw pads. Bruscino pulled the bear's lips back to reveal full rows of teeth. The bear had been healthy and simply doing what a bear will do. "It isn't that the bear was being bad. . . . At that point, he didn't know any difference between breaking a window to get food and busting apart a log," Bruscino said.[1]

At the time, there were around six hundred grizzly bears living in the Yellowstone ecosystem, a vast landscape covering some twenty-five thousand square miles, including the national park plus surrounding national forests. Often, when natural food sources were scarce, bears wandered out of the mountains and onto the valley floors where people tended to live. On a Saturday just a few weeks after no. 380 was killed, I went out to a small school between Cody and Yellowstone where volunteers were putting a tall fence around the playground to keep the grizzlies and black bears out. There'd been fifteen recent bear sightings in the area, and the locals didn't want to take any chances.

Bears and people have always had a complex, sometimes uneasy relationship, and it was no different outside of Yellowstone. The tension arose between protecting these grizzlies, which were put on the endangered species list in 1975, and protecting people and property. Chuck Neal, a retired ecologist who became a friend in Wyoming, said that too often fear and human hubris meant that bears like no. 380 got the short (read: fatal) end of the stick.

"They're more tolerant of us than we are of them," he told me in one of our first conversations.

Neal had spent his thirty-year career with the U.S. Forest Service and U.S. Department of the Interior, much of it in the West, much of it in grizzly habitat. He'd seen grizzlies more than three thousand times in the Rockies, often on solo trips "ghosting silently through the shadows of grizzly country." Over the years, he'd been charged by bears a number of times but never harmed, sure that he'd made a misstep in the bears' territory. The encounters were more exhilarating than scary, he told me, and never enough to keep him out of the woods and off the mountains.[2]

Intense and fit, Neal had no love for cities and traffic and the crush of humanity that always seemed to be imposing itself on wild places. His backyard in Cody was a fine distillation of that worldview. Where there had once been the tyranny of a uniform Kentucky bluegrass lawn, he'd made way for the wild, allowing a thicket of native bushes, plants, and trees to flourish in its place. No doubt his neighbors viewed it as an unkempt mess, but it was, indeed, meticulously laid out and fussed over. The result—a vibrant and sometimes cacophonous universe of birds, rodents, wildflowers—was hard to argue with, and he proudly gave me a tour every time I stopped by the house, sure to update me on the latest migrating bird to be spotted in the bushes.

More than anything, though, Neal obsessed over grizzly bears. He was one of the first people who walked me through some of the basics of bears in the United States, including their arrival from the Old World in the steppes of Eurasia, across the Beringia land bridge and in the New World of North America. He was the first to introduce me fully to the world's family of bears, including those who'd come before but were no longer around. *Arctodus simus* was among them.

We went into the woods together more than once. He took pity on my ignorance of so many things—how could I know the natural world without knowing the Latin names of the species we were seeing?—and viewed our conversations as an opportunity to get a more bear-centric perspective into my newspaper stories. I remember him talking to me about the giant short-faced bear as we picked our way through the crowded pines on a long walk on the steep slopes above the Shoshone River, just outside Yellowstone's east boundary. His exact words are lost, but I remember the feeling that he was, through that gravelly voice of his, revealing some sort of secret knowledge about the beasts that had been here before us. The mighty grizzly, which we'd been discussing as

I tried to keep up with him marching off-trail, would have had its hands full with the giant short-faced bear, he said, but where the grizzly and black bear survived the extinctions of megafauna at the end of the Pleistocene, *Arctodus* had not. For a time, though, this bear was "the top predator of his world," Neal later wrote.[3]

Over the ensuing years, I saw a lot of grizzly bears in and around Yellowstone. Covering the national park meant taking as many trips there as I could convince my editor to allow. I spent time with bear biologists, picked through bear scat with sticks, watched them eat and nap through sighting scopes, covered a few tourists who had been attacked, squeezed myself into an empty bear den beneath a tree, went to community meetings where the future of grizzly protections was hotly debated, wrote about famous bears who died, learned how to use bear spray, tracked the decline of grizzly food sources like Yellowstone cutthroat trout and white-bark pine seeds, and spent hours upon hours stuck in Yellowstone traffic in "bear jams" as gawkers stopped to watch a grizzly claw at the ground in search of biscuitroot and yampah. At one point, I even wrote about some stolen art in Yellowstone that had featured cheery, beer-chugging bears.

For a long time after I left the Northern Rocky Mountains, I forgot about giant short-faced bears. But at some point, I can't even say exactly when, one of them walked through a dream of mine, just a big, dim figure in the background with long legs and curved ears and a whiskered snout. It was like remembering a lost relative, the kind you heard legendary stories about as a kid and then forgot as an adult. Casually at first, I poked around for stories about the bear, nursing a small grudge that my public-school education never included a mention of it, as far as I could remember. And then I started reading scientific papers about the bears on my lunch break, printing them off at the office when no one was looking. The tug into the rabbit hole grew stronger. I stayed up late at night cruising bear and paleontological forums online and spent hours hunting down mundane details of discoveries made decades before I was born. The cave paintings of bears in France fascinated me, and so did the roles that bears play in folk tales around the world. I started hanging around obscure corners of the local university library, bothering librarians to find mis-shelved books about bears and monopolizing their copier machines with my dog-eared books and a sandwich bag of coins. I'd ride home on my bike wide-eyed with a backpack full of papers and book excerpts to eagerly devour, only to return for more a few weeks later. Before I was able to fight it off, an obsession had bloomed, the kind that journalists often contend with when they've gotten a small glimpse into a hidden world and must, at whatever cost, see the rest. It wasn't long before I was taking time off work to drive around the country, crawl into caves, and rummage through museum collections in search of bones in the same way relics are pursued by Orthodox church visitors in Russia. Somehow this bear had grabbed me around the collar and insisted that I look it straight into its long-vanished face.

I wasn't the first to fall under the spell of a bear.

8

Bears Are Everywhere

In the summer of 2014, a black bear wandered out of the woods and into the leafy suburb of Oak Ridge, New Jersey. That itself was no great excitement—several thousand black bears call the Garden State home—but this one was different. It walked upright, casually and with aplomb, through trimmed lawns and cul-de-sacs. It was never clear if he was on his hind legs because of injuries or deformities to his front legs. Whatever the case, he weighed more than three hundred pounds and seemed used to getting around on his own terms. He would've remained a local curiosity, but one of the neighbors filmed the bear and, predictably, internet stardom followed close behind. Soon he had a name—Pedals—and was starring on *Good Morning America* and phones and computers across the country. Millions saw the footage that was as delightful as it was surreal, and a little unsettling to see such a big animal ambling around hedges and standing tall near suburban doorways.

Pedals struck a nerve. In less than a year, more than three hundred thousand worried people from around the world signed a petition calling for him to be captured and relocated to a wildlife rehabilitation facility. More than twenty-five thousand dollars came in donations. State wildlife officials, though, weren't convinced that Pedals actually needed help. He seemed healthy and had survived several winters—so life in an enclosed rehab facility might've been more of a prison than a hospital, they reasoned. Pedals remained in the wild. In 2016 a bow-hunter apparently killed Pedals during the black bear hunting season. A wave of outrage ensued—both toward the hunter and the concept of hunting bears at all—and Pedals earned a memorial essay in the *New York Times Magazine*'s "Lives They Lived" feature at the end of the year alongside Antonin Scalia, Janet Reno, David Bowie, an astronaut, and the famed TV psychic known as Miss Cleo.[1]

It should have come as no surprise that a wandering bear, ordinary other than its bipedal disposition, warranted such interest. "What other animal occupies as much space in the human imagination than the bear?" asked Bernd Brunner, author of *Bears: A Brief History*.[2]

For a species that no longer has much real-life relevance to most people's lives, we sure do surround ourselves with plenty of representations of them. Professional sports, one of the far-reaching organizing forces in our entertainment culture, is rife with bears. Every major league sport has at least one bear-related mascot: the Chicago Cubs (baseball), the Chicago Bears (football), the Boston Bruins (hockey), and the Memphis Grizzlies (basketball). It's even more common in minor leagues. And by my rough count, at least twenty-five colleges have a bear as a mascot, seven more are the golden bears, another six are the bruins, another five are the grizzlies, two are the polar bears, and one is the black bears.[3]

California has a grizzly bear on its state flag, and Missouri has two (one representing strength, the other courage). Alaska's flag features the Big Dipper, a.k.a. Ursa Major, the great bear. Six states list black or grizzly bears as their official animal or mammal. America has obsessed over teddy bears, devoured stories about Pooh bear (imported from England), been lectured about wildfire safety by Smokey Bear, watched polar bears guzzle Coca-Cola in commercials, and reveled in Yogi Bear's pursuit of the picnic basket at Jellystone National Park. The U.S. Board on Geographic Names lists more than six thousand places where bears are featured in the names of towns, buildings, schools, streams, canyons, cemeteries, valleys, peaks, and other places.[4]

They've been featured on postage stamps, embossed onto license plates, and weaved into countless fables and stories, sometimes to connote wisdom and dignity and danger, other times laziness or gluttony. Real-life bears have been turned into street performers, circus acts, roadside curiosities, TV and movie stars, war casualties, and cage fighters for human amusement. Dead ones have been fashioned into rugs, coats, and even towering hats for the stiff-backed guards at Buckingham Palace. Body parts have long been harvested in the name of medicine. In *The History of Four-Footed Beasts and Serpents*, Edward Topsell in 1607 described several uses—factual considerations aside—for bears' blood, fat, and liver. Eyes too: "The right eye of a Bear dried to powder, and hung around childrens necks in a little bag, driveth away the terror of dreams."[5]

Early Scandinavian legend talks about the period shortly after the founding of the Norwegian kingdom more than a thousand years ago, when there was an army of men who possessed a sort of supernatural strength. The warriors were called "berserkers," an apparent combination of the Old Norse word for "bear," *berie*, and *serkr*, or "shirt." They fought wearing animal skins or nothing at all, some stories say, all the while locked in a savage, trance-like state. "According to ancient Germanic texts, berserkers were indistinguishable from ordinary men under normal conditions, but when aroused they could perform incredible feats such as ripping the rim of a shield with their teeth, swallowing glowing coals, or walking through blazing fire," Brunner wrote in his book. "An ordinary mortal had little chance when faced with an angry berserker, but there was a way to prevent the worst: before an outburst, a berserker would briefly become drowsy, providing a momentary opportunity to tie him up."[6]

Pliny the Elder, the Roman author and naturalist who died in the fumes of Mount Vesuvius in the year 79, offered several thoughts on bears, including these lovely nuggets: "The bear's [breath] is pestilential; no wild animal will touch things that has come in contact with its vapour, and things that it has breathed upon go bad more quickly," and "A bear's weakest part is the head. . . . The Spanish provinces believe that a bear's brain contains poison, and when bears are killed in shows their heads are burnt in the presence of a witness, on the ground that to drink the poison drives a man bear-mad."[7]

Bears show up in written texts from China thousands of years ago. They're also in the Bible and other religious writings. They've wandered into the works of William Shakespeare, Aesop, the Grimm brothers, Charles Dickens, Rudyard Kipling, and Sir Arthur Conan Doyle. They're in movies, songs, ship names, and they're still used as shorthand when the stock market has a bad day.

They have long loomed large, not just in imagination. In 1585, for instance, an explorer in the Far North named John Davis sailed into Cumberland Sound on the southeast coast of Baffin Island and spied four white figures at the foot of a mountain. "We, supposing them to bee goates or wolves, manned our boats, and went toward them: but when wee came neere the shore, wee found them to be white beares of a monstruous bignesse."[8]

In the winter of 1804–5, as Lewis and Clark's Corps of Discovery expedition spent the cold months with the Mandan people in what later became North Dakota, the Natives told stories about great and dangerous bears that lurked on the plains. Meriwether Lewis initially struck a dismissive note, figuring that the Indians' perception was formed because they had inadequate weapons. Men sharp with rifles, Lewis figured, evened the odds. The expedition got a taste of reality once they set out for the Pacific. On May 5, 1805, they ran into a five-hundred-pound bear in what is today the northeastern corner of Montana. Clark wrote: "In the evening we saw a brown or grisley beare on a sand beech. I went out with one man Geo Drewyer & killed the bear, which was verry large and a turrible looking animal, which we found verry hard to kill. We shot ten balls into him before we killed him, & 5 of those balls through his lights. This animal is the largest of the carnivorous kind I ever saw."[9]

These sorts of encounters are bound to generate stories passed through the years and assign bears an outsized place in our collective psyches. "Of the major carnivores that walk the wilderness trails of North America, three, the wolf, the mountain lion and the bear, have cast an especially strong spell on human imagination and influenced substantially the lives of those who live where they are found," Ben East wrote in his book *Bears*. "And of that somewhat mystifying trio, the bears have exercised by far the greatest influence."[10]

"We anthropomorphize many animals, though few as often as bears," Gary Brown wrote in *The Bear Almanac*. When we look hard enough, Brown notes, we can't help but see similarities. Bears stand and sometimes walk (like Pedals) bipedally. They sit on their rumps, lean back against objects to rest, scratch their backs, snore, yawn, eat sweets, leave footprints, nurse their young, discipline family members, and sometimes even sit with one leg casually crossed. We find them smart, cunning, capable, and sometimes deep and emotional, he notes.[11]

Paul Shepard and Barry Sanders go a step further in *The Sacred Paw*: "It is the bear's broad, searching, persistent openness that makes contact with us, that flash of recognition in which men instantly perceive a fellow being whose questioning provocation, whose garrulous, taciturn, lazy ways, even whose obligations and commitments to hunt, to hole up, to dominate the space he lives in are familiar."[12]

The naturalist John Muir saw a deeper, more elemental connection. "Bears are made of the same dust as we, and breathe the same winds and drink of the same waters," Muir wrote. "A bear's days are warmed by the same sun, his dwellings are over-domed by the same blue sky, and his life turns and ebbs with heart-pulsing like ours."[13]

That link, of course, dates far beyond the written word. Indigenous people around the globe have for millennia shared an important relationship with bears, one that often revolved around reverence, familial ties, and sometimes fear. Often there was a belief that people and bears were one and the same but that they simply existed in different forms. Some said bears were relatives that had wandered into the woods and became changed forever and that eating them was forbidden. Others felt that wearing the claws and fur of a bear gave them specific powers. Bears also played a prominent role in rituals around the world, including those involving healing, coming of age, hunting success, and protection. "Bears, large and powerful, have been important to people since man and bears first met," George Laycock wrote in *The Wild Bears*.[14]

The connection has been driven home time and again, including in the fight over protecting grizzly bears in the western United States in recent decades. In 2016 more than 170 tribal nations signed a treaty that opposed government-sanctioned killing policies and called for reforms combining science with ceremonial and traditional knowledge, a reflection of an ancient and revered relationship between indigenous people and the bears. "Our elders teach how the grizzly bear brought us our medicines. Grizzlies know not only about roots and herbs for physical healing but also about healing mental conditions, they say," Lynnette Grey Bull, a spokesperson of the Northern Arapaho Elders Society, was quoted in *Indian Country Today*. "The grizzly is our relative, a grandparent."[15]

David Rockwell, in his book *Giving Voice to Bear*, noted that the relationship with indigenous people in North America transcends fear and remains rooted in deep kinship:

> Bears and Indians have lived together on the continent of North America for thousands of years. Both walked the same trails, fished the same salmon streams, dug camas roots from the same fields, and year after year, harvested the same berries, seeds, and nuts. Indians came face to face with bears when both coveted the same berry patch, for instance, or when a hunter, bringing help to pack home an elk he had killed, discovered that a bear had buried the carcass and was lying on the mound. Sometimes the hunter fled, sometimes the bear. The relationship was one of mutual respect. But it went well beyond that. Bears were often central to the most basic rites of many tribes: the initiation of youths into adulthood, the sacred practices of shamanism, the healing of the sick and injured, the rites surrounding the hunt.

"For many Indians, simply speaking about bears required the observance of specific taboos," Rockwell said. He later added: "They danced as they thought bears danced, and they sang power songs to the animal."[16]

Alfred Irving Hallowell, who spent more than thirty years as a professor of anthropology at the University of Pennsylvania, plumbed the connection between Native Americans and bears in 1926 with his 175-page publication of "Bear Ceremonialism in the Northern Hemisphere."

> Many of the native tribes of North America, Asia, and Europe do exhibit toward the bear an attitude which, in contrast to that manifested toward other creatures, is more or less unique in character. Testimony to this effect is available in the accounts of explorers, travelers, and ethnologists who have sojourned with these people and there is a surprising agreement in the statements of those who have had only the most superficial contacts with the natives and of those who give us accounts based on lengthy residence and intensive study. Of course, the terms used to describe the psychological attitude of these aborigines toward the bear vary considerably. Some describe it as respect, others as reverence, veneration, or worship, but one and all are in agreement that, among the animals, bears are held in special esteem.[17]

The veneration of bears, in all its strange and wonderful forms, hasn't let up.

Once I really started noticing, it was hard to go a day or two without seeing a bear in a television commercial or on a mailbox or as a statue outside a restaurant or on a bumper sticker or a piece of jewelry or stuffed and posed for an airport display. They're an insistent and persistent presence. "Bears are everywhere," I took to saying to my wife and daughter every time I'd see one. They quickly grew tired of my mantra repeated over the course of a year or more, sometimes replying: "So we've heard."

On one driving trip through southern Oregon and Northern California, there were times when it seemed as if I couldn't go more than a few minutes without seeing some facsimile of a bear. In Grants Pass, Oregon, bear statues lined the main drag. Some were in outlandish costumes—farmers, artists, loggers, bakers, and even something akin to Carmen Miranda—and others were posed serenely, just off the sidewalk and hidden around corners. A local bank started BearFest in 2003 by commissioning twenty life-size fiberglass bears. Since then, nearly two hundred more have been created as public artwork and items to be auctioned for charity. Travel south on the Redwood Highway and every small town has a wood carver selling wares on the side of the road. Inevitably, bears are the most prominent creatures on display, typically perched on their hindlegs tall and proud. Farther south still, a bridge that crosses the Klamath River is guarded at each end by two giant metal-cast bears, plump on all fours and painted as golden and as brilliant as the sun. "Bears are everywhere," I murmured as the bears disappeared in my rearview mirror. "Bears are everywhere."

But not *all* bears are everywhere. Largely absent in most of the blizzard of bears in our culture is the North American giant short-faced bear. It may have been the biggest bear ever to walk the continent, but it seems to have left hardly any visible imprint on our art, culture, and collective consciousness. It has no sports team named after it, no festival, no iconic cartoon or indelible place in our collective mind. The more I obsessed over the short-faced bear, the more I remained perplexed that hardly anyone knew about it. Despite our deep connection with bears around the world, there seemed to be barely any lasting memory of this particular bear's time on stage here in North America.

PART 2

Goo

9

La Brea

If you're looking to peer straight into a Pleistocene nightmare of death and suffering—and somehow I was—you could do a lot worse than the La Brea Tar Pits in downtown Los Angeles. They are there, lurking just off sunny Wilshire Boulevard surrounded by glassy high-rises, Starbucks, cafés, billboard-sized movie posters, city buses with hissing brakes, honking cars, tony homes with colorful front doors, and, often, a construction crane casting a long shadow. No matter the fancy modern trappings, there's no escaping the dark history of this place. Walk around and it's bubbling up beneath your feet.

The black goo is asphalt from a natural oil field a thousand feet below that's been forced to the surface through cracks in the rock beds atop it. What finds its way to daylight is the texture of ultra-thick, impossibly sticky peanut butter that smells like the hot goop that's put on roads. The stench is the least deadly aspect, though. In its heyday tens of thousands of years ago, a foot or even a few inches of the stuff was enough to ensnare the biggest and mightiest of beasts. From there, the horrors just got worse. Any trapped animal would've struggled and made noise. That in turn would attract others, especially flesh-eaters on the hunt for an easy meal. On approach, they too would often get stuck and begin to scrap and cry, attracting the attention of still more hungry visitors willing to step into the smelly muck on the chance of an easy morsel or two. By the time they realized their mistake, it was too late. Fighting the gooey asphalt led to still more frustration and exhaustion. Futility and fear thickened in the air. Those that didn't die of predatory wounds—like a torn-off limb or deep slash to the throat—succumbed to thirst and starvation and eventually disappeared into the muck to make way for the next victims. The slow-motion dance of death lasted some fifty thousand years.

Something else remarkable happened. The tar was perfect for holding bones in a kind of suspended animation that kept them preserved and well hidden for tens of thousands of years. Today, scientists at the La Brea Tar Pits and Museum estimate more than 3.5 million specimens—bones, teeth, skulls—have been pulled from the goo. It's a macabre and astonishing portrait of who lived here at the time. More than 600 types of animals and plants have been found, including mammoths, mastodons, sloths, and bison plus about 140 bird species, 158 types of plants, 3 types of birds, and 25 families of beetles.[1]

Most notable, and most famous, are the big predators. More than 2,100 individual saber-toothed cats have been identified, along with the remains of about 1,600 dire wolves. It's also the largest assemblage of giant short-faced bears ever found. So far, parts of more than thirty individuals have been pulled from the sludge, ambassadors of another time that's both far away and incredibly close.[2]

My first trip to La Brea was a sunny Friday in the early spring. It felt like every first grader in Los Angeles County was also visiting. There was an anthill-like quality to the frenzy of young humanity in the plaza outside the main entrance. Behind them, just off Wilshire, was the fenced-in iconic Lake Pit featuring life-size fiberglass statues of a family of mammoths in the throes of sticky crisis—an adult half-submerged and keening while a young mammoth and another adult watch in distress from the safety of the shoreline. The pit is actually a manmade leftover from an old asphalt mining operation, but the point is well taken: this is a place of death and otherworldly spectacle.

Inside the main building the cacophony of young voices only intensified. Still, it was heartening to see so many kids fawning over extinct animals that weren't dinosaurs. Not that I have anything against dinosaurs, but they've had their moment, haven't they? It's clear who the stars are at La Brea. Mammoths and saber-toothed cats dominate the museum's iconography, and not without reason. Both push the imagination into new places that only monsters can. A lion-sized cat with an extraordinary set of over-size canines that singularly embody nature at its most ferocious? Yes, please. A building-size member of the elephant family with wondrous curving tusks that actually stood where I'm standing? Count me in.

Nonetheless, just past the main doors, I made a beeline to the heart of the museum, where there's a towering, life-like statue of *Arctodus simus*. It's standing on its hind legs, knees slightly bent, lanky arms ready to grasp anything close, and jaws wide open to showcase a hall of meat-tearing teeth and grinders. The feet and paws are massive, and its face, gazing into the distance, seems to insist on its utter dominance over all living creatures. I stared up from below and imagined the stomach-churning noises that came out of its mouth, probably the last sound searing the brain of many an animal on its way to the other side of life. The bear statue, perched on a foot-tall stand made to look like a rock, probably reaches fifteen feet tall. It's an exaggeration of the size of these bears, but it's a magnet for adults and kids, all wanting to have their photo taken in its presence. A more realistically sized skeleton of a short-faced bear is just a few feet away, demure on all four feet, but it doesn't attract the same kind of attention.

Behind the giant bear statue is an animatronic giant sloth that's having its neck gored by a saber-toothed cat on its back. Again and again the *Smilodon* plunges its cone-shaped teeth toward the top of its neck. The attack comes with a robotic roar, but the kill is never quite consummated. I watched this faux killing for several long minutes alongside a crowd of kids spellbound by the death dance being reenacted both for our pleasure and edification.

I had to really look for other signs of the short-faced bear at the museum. Not far from the statue, just past the wonderful wall of 404 left-facing dire wolf skulls silhouetted against an amber light, I spotted the bear in a large-scale mural of the Pleistocene. It was standing on its hind legs, arms hanging loosely at its sides, watching a herd of bison. It was alone and had a sad, human-like quality. Outside of the main building, on the grounds northwest of the museum, there was another short-faced bear statue. This one, commissioned in 1920, was slightly more cartoonish, a dumpier version sitting atop a rock with one arm casually draped over its knee and mouth slightly ajar, like it's ready to have a smoke and shoot the breeze. He seemed lovable in a bowling buddy sort of way, all soft edges and not at all worried about sitting smack in the middle of the Pleistocene's greediest and most toxic death trap.

Before I left, I went back inside to peer into a window leading to the staff-only collections room. My heart sped up for a second with the sight of a stack of gray pull-out trays hand-labeled in black marker with "Arctodus." I'd have to come back for those.

10

Hiding in the Muck

As far as I can tell, there's only one person who's been fully submerged for an extended time in the La Brea Tar Pits and lived to tell the tale. And, true to form for La Brea, the story has its own dark side.

Just after 2:00 a.m. on May 13, 2011, a Los Angeles nightclub owner named Alonzo Ester steered his white three-hundred-thousand-dollar Rolls Royce into the driveway outside his mansion in a tony neighborhood called Baldwin Hills. Moments later, while he was still in the driver's seat, he was shot multiple times. Ester, sixty-seven, died less than an hour later at the hospital.[1]

As investigative leads dried up, police put out a call for tips along with a fifty-thousand-dollar reward. A call eventually came in that a gun used in the murder—or somehow connected with the crime—had found its way into the main Lake Pit outside the Page Museum, which is within easy range of an over-the-fence throw from Wilshire Boulevard. Who would ever find it there? Search crews checked out the tip and, sure enough, using sonar from the surface, spotted what they thought might be the weapon. They tried to fish it out with a long rod and magnets and other tools, all to no avail. A diver started to wade in wearing protective equipment and immediately got stuck and had to be cut out of his wetsuit to escape. The operation nearly ended there, but by then the media had been alerted and some of the top brass at the LAPD asked for a volunteer to plunge into the black morass and fetch the weapon. If nothing else, it was a bit of good publicity: tenacious cops going to every length to solve a case, even disappearing into one of the city's deadliest landmarks. Who was ready to descend into a deep pit of toxic tar that was so sticky that it once fatally trapped mammoths, saber-toothed cats, giant bears, and all manner of trespassers?[2]

That's how, on a sunny summer day in June 2013, a Los Angeles Police Department diver named David Mascarenas purposefully sank into black muck in search of a gun. As soon as I heard the story, I knew I wanted to talk to him. Short of dipping into the oily goo myself, this was as close as I was going to get to a firstperson account of what it's like to get trapped in the La Brea Tar Pits, and what it might've been like for my bears. It took us several months to connect but I finally reached him on his cell after work late one afternoon. Even though it had been six years since his adventure at La Brea, he spun out the story in the clipped, nononsense way that some cops do. As the officer in charge at the scene and a diver since the early 1980s, he volunteered to go into the tar, albeit with a bit of a bad feeling in his gut.

"I've done a lot of crazy stuff for this department: bombs, bodies, narcotics, going under boats, bailing people out," he told me. "I try to lead by example. And I thought, 'Hey, if something happens, it's a good way to go out.'"

Crews at the scene ran a couple guy wires across the lake and greased the bottom of a shallow jon boat so it could be more easily pulled across the pit. Meanwhile, Mascarenas geared up with a diving suit, thick gloves, helmet, breathing apparatus, an air supply line that would go to the surface, and a harness that wrapped around his torso. Every seam on his suit was covered in duct tape. Given enough time, the corrosive goo would eat through any weak point. By the time he was ready to go in, a crowd had gathered and TV cameras were rolling. Gingerly Mascarenas lowered himself into the lake and descended through layers of tar mixed with water and pierced with bubbling columns of methane. He described it as dropping through sludgy, green-black pudding. With virtually no visibility, he used a long pole to pull himself deep into the muck, adjusting his course as the crew watching the sonar on the surface radioed directions into his ear. His depth gauge stopped working at seventeen feet and time slowed to a crawl.

"It was a very eerie feeling like you're in space and all by yourself," Mascarenas said. "I actually got stuck twice and didn't think I was coming back out."

At one point, the crew pulled on his harness so hard to free him he thought his ribs might break. But they also told him he was close to the weapon and encouraged him to keep going down. Extraordinarily, while his fins and gloves began to fray from the heat and toxic chemicals and his other equipment started to fail, Mascarenas found what he was looking for and returned to the air and Los Angeles sunshine. He was lightheaded, coated in goo from head to toe, relieved and a little shocked that it had actually worked. He'd spent seventy-seven minutes submerged in the tar. The recovered gun helped secure a plea deal from one of the suspects.

"That was a one-off caper," said Mascarenas, who was a few months from retiring when we talked. "It's never going to happen again."

Technically, the tar pits aren't tar. The shiny black liquid that permeates the ground is actually asphalt, a natural bituminous substance that begins its life far underground. "Tar" is an artificial substance that's made from the distillation of things like coal or wood. Nonetheless, "tar" is what it's been called for years, and the name has stuck (even the museum and grounds are officially called La Brea Tar Pits and Museum).

When I was there, several neon green cones were scattered on the park grass around the museum. Each had "Sticky" written on it and was placed beside a small fissure in the ground where the tar had imposed itself on the surface. Once, I dipped a finger and spent the rest of the day trying to get the goo off and wondering how it had subsequently found its way onto my shirt, pants, notebook, and backpack.

Others have put it to much better use. Native people used the tar for thousands of years as a waterproof sealant for boats and baskets. It was a superb adhesive and building material.[3]

The first written record of the pits seems to be from Gaspar de Portolá's notes of his California expedition. On August 3, 1769: "We proceeded for three hours on a good road; to the right of it were extensive swamps of bitumen which is called *chapapote*. We debated whether this substance, which flows melted from underneath the earth, could occasion so many earthquakes."[4]

There was another report in 1792 by explorer José Longinos Martínez:

> Near the Pueblo de Los Angeles [there are] more than twenty springs of liquid petroleum, pitch, etc. To the west of the said town, in the middle of a great plain of more than fifteen leagues in circumference, there is a large lake of pitch, with many pools in which bubbles or blisters are constantly forming and exploding. . . . In hot weather, animals have been seen to sink into it, unable to free themselves because their feet were stuck, and the lake swallowed them. After many years their bones come up through holes, as if petrified. . . .
>
> For a great distance around these volcanoes there is no water, and when the heat of the sun forces birds to seek it they alight upon the lake, mistaking it for water. All the birds that do so are caught by the feet and wings until they die of hunger and thirst. Rabbits, squirrels, and other animals are deceived in the same way, and for this reason the gentiles keep a careful watch at such places in order to hunt without effort.[5]

The land eventually became known as Rancho La Brea, and 4,400 acres were given to Antonio Jose Rocha as part of the Mexican land grant, with the proviso that the locals could continue using as much asphalt as they wanted. Over the years it was subdivided and developed, and eventually fell into the hands of George Allan Hancock.[6]

The state geologist described the site like this in 1865: "A very large amount of the hardened asphaltum, mixed with sand and the bones of cattle and birds which have become entangled in it lies scattered over the plain." Within a few years, entrepreneurs were capitalizing on the sticky material. Giant pits were dug by hand, and thousands of tons were removed for paving and roofing. By the mid-1870s, vast quantities were being boiled on site, set into molds, and shipped to San Francisco for twenty dollars a ton. "The supply is almost inexhaustible," one report said. But there was more than just pitch and modern bones. In 1875 Hancock gave geologist William Denton something incredible: a nine-inch curving tooth with a broken crown and a base of nearly two inches across. It had been found in a pit about fifteen feet below the surface. This wasn't anything from the modern faunal suite. Indeed not. Denton quickly surmised it was the prehistoric tooth of "a great sabertoothed feline" larger than any of those that had been found in Europe.[7]

Although his findings and theory were featured later that year in the *Proceedings of the Boston Society of Natural History*, they were quickly forgotten, and the work in Los Angeles of digging up and selling the asphalt continued.[8]

A geologist named W. W. Orcutt came to the ranch in 1901 to investigate a failing oil well not far from the tar pits. He was enamored with a small mosaic of fossils and pitch he found: "The white weathered bones made a beautiful contrast with the black surface of the hard asphaltum." He returned several times over the next five years, collecting what turned out to be part of a saber-toothed skull, the jaw of a dire wolf, and remains of a giant ground sloth.[9]

The bones eventually found their way to John C. Merriam at the University of California, who had played such a key role sorting the prehistoric bones from Potter Creek Cave, including the giant short-faced bear. He came immediately to Los Angeles and found, in the open ground between two sets of oil derricks, a series of ponds, pools and puddles of black tar, some warm and squishy, others hardened and tough. There was a particularly large quagmire near the middle of the property. "The water of the pond is heavily loaded with oil and tar, and through it great gas bubbles several feet in diameter are constantly rising with a loud splash," Merriam said. And everywhere, he said, there was a "great freight of bones." They were so fresh-looking and near the surface that it made sense that they'd been dismissed as merely the ho-hum remains of cows, horses, dogs, and other common domesticated animals. A closer look, though, revealed many to be "strange, extinct animals which lived in an earlier geological period." Soon Merriam realized he was standing in "one of the most remarkable accumulations of prehistoric remains in the world."[10]

He quickly sought permission for a wide-scale fossil excavation operation on the property. The years-long project that followed would change the course of Merriam's career and the modern understanding of the animal lives that had come before.

Although word about the extent of the discovery found its way into scientific circles, the public's detailed knowledge of La Brea didn't really begin until the October 1908 issue of *Sunset* magazine, whose cover featured a colorful if primitive painting of a saber-toothed cat parting some tall grass, its jaw open and eyes hungry for prey: "The Sabertooth Tiger in California—World-Old Found in Asphalt." Inside was Merriam's account of the find with the headline "Death Trap of the Ages." His ten pages of plainspoken prose from Rancho La Brea made for fascinating reading. He described in detail the fossilized inhabitants of the gummy pits. Some were ordinary— like ducks, dogs, cats, owls, squirrels, beetles, and centipedes—and others extraordinary. There were great extinct wolves, previously only known from Indiana and Texas, alongside "powerfully but clumsily built" giant sloths as well as ancient bison, camels, and mammoths. There were also bears and saber-toothed cats. In one spot less than two yards square, eighteen complete saber-toothed skulls were found along with a complete skeleton. This field of death fired Merriam's imagination with scenes of horror and fascination: "The peculiar ducks and pelicans and condors, the young cam-els, bison, horses, and deer, with the mammoth and the ground-sloth, have sunk in the pitch, and in their struggles have enticed the wolves, bears and the sabre-tooth cats. Sometimes a single struggling animal may have attracted several wolves or tigers, and around its body a combat was carried on which ended in both the victor and the vanquished being swallowed up in the tar."[11]

The story had only been partly told, *Sunset's* editors told readers in a note above Merriam's piece. "Strange it is to note that the Death Trap is still set. Every day the treacherous pool catches a crane or duck or smaller bird, holding them tight in sticky grasp, and laying them away for the wonder of some paleontologist of the unguessed future." Scientists, the editors said, were still "exploring the black secrets of this almost dateless prison."

What followed was one of the most fruitful paleontological digs ever in the United States. Between 1913 and 1915 alone, about five hundred thousand fossils were pulled out of the muck by scientists, professional diggers, students, and volunteers. Nothing like it had ever been seen before. The granular asphalt was perfect for preserving bones, stripping away the soft tissues and leaving behind the harder parts like skulls, teeth and bones. A single chunk of less than four cubic yards yielded portions of more than fifty dire wolf skulls and at least thirty saber-toothed cat skulls, along with bones of bison, coyotes, sloths, and birds.[12]

The bones were spread irregularly but often ended up in pockets with clusters of animal remains piled on top of each other. It's not hard to imagine how catastrophe unfolded, especially when the tar was covered in a layer of water. Here's how Finnish paleontologist Björn Kurtén imagined it:

> Animals come to drink, get caught, and struggle to break away. The struggle promptly catches the interest of every flesh-eating animal in the neighborhood. The fossil-collecting trap is baited. Here comes the dire wolf, the hyena of the New World, the hungriest of them all; and he gets caught and perishes. Here comes the monster sabertooth, and he gets caught too. Soaring far away, condors and vultures sense the commotion and swoop in; many are added to the bag. Insatiable, the tar pool pulls its victims down, one after the other, strong or feeble, in a chain-reaction which ends only when there is no meat-eater left nearby to succumb to the lure.[13]

While the carnage count is high, it accumulated over time. "It is conceivable," says one of the most definitive early histories of La Brea, "that a single seep might gather into its mass in a relatively short time a great many victims whose remains now form the remarkable accumulation of bones, skulls and teeth found at Rancho La Brea. Nevertheless, because of the 30,000 or more years represented by the fossil accumulations, only one such major entrapment episode (involving, for example, a large herbivore, four dire wolves, a sabertooth, and a coyote) need have occurred every decade to account for the immense number of fossil animals represented in the collections."[14]

The early digs at La Brea were difficult affairs, a grueling combination of sweat, filth, hard labor, primitive tools, and the occasional lighting of kerosene (not advised) in the hopes of burning away the oily coating around fossilized treasures. It was an extraordinary flurry of activity that laid the groundwork for the eventual donation of the land to Los Angeles County, formation of Hancock Park and the opening of George C. Page Museum in 1977. Excavations continue today, albeit at a less frenzied pace and in far fewer areas.[15]

It's still not easy. Since the position of each fossil is crucial to understanding the context of what happened when it died, each excavation site is divided into square three-foot grids, which are in turn subdivided into six-inch grids. Digging commences and each fossil find is carefully noted in field notebooks with sketches and photos. The large fossils are then pulled out, and the mixture that's left—the matrix—is put into screened baskets and boiled in solvent to remove the asphalt. Degreasers and a solvent are used to further reduce the mixture to sand, small rocks, and tiny "microfossils." Those are then sorted by hand with camelhair brushes under lighted magnifiers—insects, plants, wood fragments, and seeds along with other tantalizing bits like the tooth of a mouse, the vertebra of a snake, or the rib of a fish. It all gets saved, categorized, and stored. It's a fascinating process to watch, and the museum indulges with a large, glass-enclosed room where visitors can watch the entire thing, not unlike watching industrious animals at the zoo, although these specimens wear white lab coats and pretend not to notice the crowds a few feet away.

The bigger bones, of course, get special treatment. Some are lowered into jars of solvent and then put into ultrasonic tanks where sound waves help the solvent separate the specimen from the asphalt. Others are cleaned by hand with tools and solvents. A "time-consuming and dirty job," the museum says. But the resulting clean fossil can then be properly identified, usually by comparing it with the skeletons of known specimens in the science lab. Some of them eventually end up on display, but most find their way to a shelf or drawer in a back room—tagged, sorted, and waiting for someone to tell their story.

11

Teeth and Bones

Aisling Farrell and Alexis Mychajliw were ready for me when I went back to La Brea later in the year. I was ushered through the staff entrance at the back of the building, stopped at a security desk and then escorted to a back room of the museum, the science lab. The main room has a low ceiling with florescent lights that light up row after row of drawers and cupboards full of fragments pulled from the pits. But even those aren't enough. Like an overflowing basement, many of the tops of cabinets and tables were populated with bones, casts, skeletons, and miniature models of species at La Brea: dire wolves, dwarf pronghorns, mammoths, saber-toothed cats, bison. They seemed to have a life of their own, staring back at us from a different time.

Farrell and Mychajliw knew I was there for the bears and kindly accommodated me. Spread out on a work table were parts of a giant short-faced bear, the mandible of a grizzly, the skull of La Brea's only black bear, and the skull of a Kodiak brown bear—obviously not from La Brea, but it was helpful to see the size comparisons. Behind us were several gray trays on a rack with more *Arctodus* fossils, including teeth, jaws, vertebrae, and leg bones. Most of the giant short-faced bears in the collection were removed during the intensive excavations in the early 1900s, but a few others have emerged over the years. The last *Arctodus* fossil likely came out in the 1980s.

While we chatted, there were two items I couldn't stop looking at. One was the dark-brown jaw of a short-faced bear about ten inches long with two or three molars still attached and a great, arching canine with a worn-down tip. The other was the top half of a huge *Arctodus* skull that was blackened and beat up. Half of it was an artificial cast, it turns out, but well done. The thick nasal cavity, wide eye sockets, and a few remnant teeth made it impossible not to imagine the brutish damage that its owner did at one time. "They were impressive," smiled Farrell, who manages La Brea's collections.

Although more giant short-faced bears have been found at La Brea than anywhere else on Earth they only came from a handful of pits—and they're far, far outnumbered by other big predators, including saber-toothed cats and dire wolves. Curiously there's no evidence that giant short-faced bears, the biggest meat-eaters around at the time, directly competed for prey with other predators at La Brea. So what did they eat?

Farrell was one of eight scientists, including others from the United States, Spain, and Brazil, who coauthored a 2017 study to get at that question. It's an important one for understanding how they fueled their magnificently large bodies and interacted with other big beasts. The debate has revolved around a few key possibilities: that the bears were "super predators" that mostly feasted on the flesh they killed, that they were opportunistic scavengers of already-dead carrion, that they had omnivorous tastes that included plants and animals, or even that they were strict vegetarians like their distant cousin, the giant panda.

Teeth, the hardest parts of the body, are one of the most obvious places to look for clues. They're a dead giveaway, whether they're flat, broad teeth that are perfect for grinding plants or jagged incisors that can rip flesh. They also act as a sort of secret diary for scientists searching for hints about nutrition, the local environment, and whether they were living with aches and pains.

For the 2017 study, researchers closely examined thirty-three molars from giant short-faced bears from La Brea, plus others from Canada and Alaska. They found that the La Brea bears were commonly afflicted with cavities—"carious lesions" is the technical phrase—that come with eating sugary carbohydrates. Turns out, the bears were regularly eating plants, possibly juniper berries and honey, both of which have been found in fossil form at La Brea. Without the services of a dentist, the sugars were rotting their teeth. The same couldn't be said for the bears in the Far North, which had long been suspected of diets that more heavily relied on meat.[1]

The study added an important wrinkle in the story of these giant bears. They were omnivores just like today's black bears and grizzlies, whose diets vary greatly depending on where they live. Like those bears, giant short-faced bears were almost certainly adaptive when it came to eating. They made do with what was available, eating flesh when possible, switching to plants when needed or preferred. That not only helped ensure they could live in different kinds of ecosystems and seasonal variations, but it also may explain why they were one of the last big animals to go extinct at the end of the Pleistocene. When the world is going to pot (or at least undergoing rapid change), those who stubbornly hold on to the old ways are apt to get swept away. Adjust and you just might survive.

The teeth study revealed just a tiny aspect of the nature of La Brea's bears. There are still other, larger questions unanswered, including why parts of more than thirty giant short-faced bears have been unearthed there while just a handful of grizzlies and a single black bear have been found. And why were there so many more saber-toothed cats (two thousand or so) and dire wolves (a thousand-plus)? Plus, come to think of it, why do large predators dominate the sheer number of bones found at La Brea, especially knowing that prey species necessarily outnumber predator species in every ecosystem?

Bears are often solitary figures, and that might explain why so few, relatively, have been found at La Brea. And it makes sense that the cries of a single distressed mammoth or deer might draw an abnormal number of hungry predators attracted by the prospect of an easy meal. Murkier, though, are the detailed interactions between dire wolves, saber-toothed cats, and giant short-faced bears. Did one group dominate? Did the presence of a single, freakishly large bear change the dynamic on a given day? Did the bears steer clear when other big predators were having their way?

"This was a Serengeti-type environment with lots of herbivores, but it was also dense with carnivores," Farrell told me. "There's still a lot to learn about their behavior."

After a while, Mychajliw and I stepped out of the lab, walked through the museum buzzing with visitors, and into the sunlit park outside. I wanted to see famed pit 91 on the west side of the property.

During the Pleistocene it was little more than a shallow pool of sticky asphalt that captured those who came too close. Over time the dirt and sand that washed into the pool was buried underground along with all the bones it had collected over thousands of years. Although it was the ninety-first hole dug between the 1913 and 1915 excavations, it quickly became one of the most fruitful, famous, and important. Thousands of fossils have been pulled out of pit 91, and scientists estimate it's responsible for doubling the number of species known at La Brea. Volunteer excavators still work at pit 91 each summer. So far, they've dug down 15 feet, identified fossils ranging from 44,000 to 14,000 calendar years ago, and found incomplete specimens of 73 saber-toothed cats, 56 dire wolves, 16 coyotes, 13 horses, 12 bison, 6 ground sloths, 6 giant jaguars, 4 short-faced bears, 2 camels, and a mastodon.[2] That says nothing of the plants, bird beaks, fish bones, and insect exoskeletons that have been retrieved. Or the 45,000 bits of mollusks. It is, the museum claims, "one of the world's longest-running urban paleontological excavation sites."[3]

Mychajliw took me through the locked door of the outside display and onto a platform overlooking the pit. It was about twelve feet by twelve feet and surrounded by wooden platforms. Pink string tied above the gooey black pond divided it neatly into a series of squares, and there were a series of plastic flags stuck into muck where a new fossil had been found. One of the volunteers was busy siphoning off the top layer of liquid in one of the grids.

Not all of the giant short-faced bear specimens from La Brea have been dated, and some may never be, Mychajliw said, but one of them from pit 91 has been. It was around twenty-eight thousand years old. Not far away, she pointed out another pit where two grizzlies, an adult and a juvenile, were removed more than a century ago. Those bears are important players in a broader story that Mychajliw was trying to piece together, one that had particular relevance at the moment.

Mychajliw is part of a multidisciplinary project called the California Grizzly Research Network looking at the potential for returning grizzly bears to the Golden State. California was once home to as many as ten thousand grizzly bears before the 1849 gold rush, from the High Sierra in Northern California to the Mojave Desert, Los Angeles, and the state's beaches, forests, and valleys. An ugly frenzy of unregulated trapping, killing, and poisoning after 1849 drove the bear into oblivion. It took just seventy-five years to send California's grizzlies into the maw of extinction. The last credible sighting was in 1924, but the grizzly remains on the California flag and is one of the state's most enduring pieces of iconography.[4]

Discussion had been heating up for several years about whether to reintroduce grizzly bears to parts of California. The California Grizzly Research Network formed in 2016 to inform that debate with rigorous research. Mychajliw, a postdoctoral fellow at La Brea who often works at the intersection of ecology and paleontology, was on a team looking at the ancient distribution of bears in California, including black bears, grizzly bears, and giant short-faced bears. La Brea's rich fossil record, as well as bones held in museums, colleges, and private collections, might provide a unique window into whether—and how—all three bear species may have coexisted in California. There could be other clues about what they ate, where they preferred to live, and how they responded to changes in their environment. Giant short-faced bears and black bears lived for thousands of years in North America before grizzly bears came along (around the same time people arrived). Today only the black bear survives in California. What might happen if another big bruin returns?

For months Mychajliw had been poring over bear bones from around the state, removing a fingernail-sized sample from each to be processed and analyzed. At a lab at the University of California—Irvine, the samples were demineralized and the resulting sponge-like collagen freeze-dried. She then picked apart the sample—now the consistency of cotton candy—to analyze carbon-to-nitrogen ratios that indicate what the bear may have eaten. That material is later super-heated into a gas that can be used to determine how long ago the bear lived on the Earth. When I visited, Mychajliw was not-so-patiently waiting on her results to come back.

In the same fenced-off area that houses pit 91, she showed me one of the open wooden boxes packed with a matrix of hardened tar and animal bones. It was awaiting the painstaking process of sorting, categorizing, and cleaning. There may be bear bones locked somewhere in there—giant short-faced, grizzly, or otherwise. Like any sample, it may or may not be viable, but every bit gives a better picture of the lives of these bears thousands of years ago. "That's really what I care about right now," she said.

Back in the La Brea lab among the skulls, jaws, and leg bones, the talk turned to something more philosophical. Handling the bones of individual bears that lived hundreds or thousands of years ago can be humbling and even a little strange, especially when you remember that each was connected to a body that had a rich life and a brain that puzzled over things like giant sloths and saber-toothed cats, and that each had a sophisticated system of surviving in a world that might seem slightly foreign and exceedingly dangerous to us today. For the most part, Mychajliw was tightly focused on the detailed scientific work in front of her. "Sometimes, though, you have to remind yourself that this tiny crumb of bone is from this giant thing," she said.

On my way out of the Page Museum, I stopped at the glassed-walled, U-shaped gift shop, curious to see how the giant short-faced bear was presented for souvenir hunters. Predictably, saber-toothed cats and mammoths played a starring role on T-shirts, mugs, hats, and toys. Second place went to sloths and wolves. The only nod to *Arctodus* I could find was on a side wall: a little package of short-faced gummy bears that I bought and tossed into my backpack.

After my visit to La Brea, I decided not to stay the night in Los Angeles. By then, I was sick of the traffic and the crush of people at every turn. Even though it was late in the afternoon, and against my better judgment, I decided to drive the nine hours back to Tucson and be done with it.

The traffic on eastbound Interstate 10 was stop-and-go-and-stop for more than two mind-numbing hours. Somewhere outside the city, I spotted an extremely large California state flag above a highway gas station, white with the red star at the top left, red bar along the bottom, and, smack in the middle, a grizzly bear wandering through the scene. My mind turned to the story of Monarch, the bear who served as a model for the flag. It's a pretty fitting example of the dual nature of our perverse relationship with bears, one where awe and fear seem to be locked in an endless battle. The story goes that during the prolonged grizzly killings of the late 1800s, newspaper mogul William Randolph Hearst hatched a publicity stunt to bring a live California grizzly to San Francisco. He dispatched one of his journalists to complete the job—and the reporter returned months later with a 1,100-pound bear from the Ventura Mountains. The bear was named Monarch (Hearst's *San Francisco Examiner* sported the "Monarch of the Dailies" tagline) and held in dismal captivity for the next twenty-two years while the last of California's grizzlies were pushed over the edge of extinction. After Monarch died in 1911, his taxidermied body was used as a model for the California flag. Today his pelt is at the California Academy of Sciences in San Francisco. The rest of his body was buried and later exhumed, and now his skeleton is stored at the University of California—Berkeley (likely not all that far from the giant short-faced bears pulled from Potter Creek Cave).[5]

In life, Monarch was purposefully separated from his wild home, and in death his bones were separated from his body. It's a peculiar kind of cruelty that only humans seem capable of. No wonder the *San Francisco Chronicle* dubbed him a "symbol of suffering" in a centennial story about the flag.[6]

I had to wonder: If giant short-faced bears had survived the end of the Pleistocene, would Monarch have been spared so that an even bigger bear could be on the flag? Certainly someone like Hearst wouldn't have settled for something so pedestrian as a grizzly when a much, much bigger bear was available for the taking. Maybe it wasn't so bad that *Arctodus simus* didn't make it this far. And what would life have been like anyway? So much of Southern California is a never-ending circus of streets, highways, strip malls, office parks, airports, and sparse pockets of "nature" that are as manicured as they are deficient in wildness. If it's hell enough for a person, how might it be for the mightiest bear that ever walked the continent?

Later on the drive, past Indio and before Blythe, the interstate emptied out and the sun put the day's final coats of bruise-colored light on the Mojave Desert. Dead ahead, a small range of mountains emerged on the far horizon. One of them looked exactly like the lower jaws of a bear that I'd been handling back in the science lab at La Brea. The flat molars in the middle and back were followed by the beautiful half-crescent of a canine jutting madly toward the sky.

12

A Surge of Discovery

While Merriam and others in Los Angeles were sorting through the tar pits of La Brea, miners some three thousand miles to the north in the Yukon Territory were looking for gold. About thirty miles southeast of Dawson—which had been the hub of the Klondike gold rush in the late 1890s—crews were digging near Gold Run Creek. About forty feet below the surface, they found an enormous animal skull minus its lower jaw. In May 1911 one of Canada's most famous paleontologists at the time, Lawrence M. Lambe, published a paper in the *Ottawa Naturalist* laying claim to the largest version of the bear yet. Despite its short muzzle, the skull was more than twenty inches long, much bigger than the specimens from Potter Creek Cave. The head was exceedingly broad, the forehead high, the teeth "much worn," and the canine protruding out of its socket measured more than two inches long. It "represents an individual of great physical power and bulk," said Lambe, who tended to traffic in dinosaurs rather than Ice Age beasts but knew a monster when he saw one.[1]

Although it seemed to fit into the *Arctotherium* family, as it was known then, the bear was so large and distinct that Lambe suggested it receive its own name: *Arctotherium yukonense*. His paper, complete with photos of the skull, was an important leap forward in understanding this still-mysterious bear. Now scientists knew it was capable of growing to great size and that it had once been in the far northern reaches of the continent. That region of the Yukon, which remained unglaciated during the Pleistocene, would later reveal a trove of Ice Age fossils, including the upper arm bone of a giant short-faced bear found in 1973 that lived an estimated 29,600 years ago.[2]

There were other discoveries.

In April 1907 a crew working for the American Lime and Stone Company in central Pennsylvania blasted into a cave that was about forty feet long and seven feet wide. The floor was covered with about two feet of red dirt, along with limestone blocks and hardened bits of cave onyx. The mixture also held the bones of many, including bison, wolves, musk ox, sloths, mastodons, and what would years later be identified as a giant short-faced bear. The Carnegie Museum of Pittsburgh sent O. A. Peterson of its paleontology division to collect the bones, but he had to work fast. Some of the material was already being whisked away by "curio hunters" and other bones were simply finding their way to the dump piles. Over several weeks, there were no complete skeletons to be found but plenty of scattered bones. "As is usual in all limestone caverns, the material was found much disturbed and mutilated by fallen blocks, large and small, which dropped from the roof and sides of the cave," Peterson wrote later in the *Annals of the Carnegie Museum*.[3]

Scientists at the Pennsylvania Geologic Survey speculated that most of the animals fell into Frankstown Cave, as it was eventually named, through a hole at the surface that was later clogged and filled. It raised a curious question, according to the agency's 1930 report: "Bones of one adult and five small mastodons were found in the upper part of the deposit, as if they had been the last to enter the cave. Snakes and frogs are always falling into holes, bats live in caves and one can imagine a peccary falling in and a wolf going in after it, but it is hard to understand how six mastodons were so stupid as to be thus trapped."[4]

As for the bear, researchers were able to find teeth, arm bones, leg bones, and vertebrae of at least two individuals.

Farther west, in 1916, the arm bone of "a gigantic bear" was found in Cass County, Nebraska, and was temporarily dubbed *Dinarctotherium merriami*.[5] Two years later an ankle bone from a "very large" ursid, possibly something from the *Arctotherium* family, was found in northwest Nebraska along the Niobrara River.[6]

In 1921 crews building a new highway in California's San Joaquin Valley came across a fossil-packed bed of black sticky asphalt. In the years that followed, pieces of hundreds of well-preserved Pleistocene animals were unearthed from the sprawling McKittrick Tar Pits, including insects, birds, plants, red deer, the American lion, and a nearly complete skull and jaw of a giant short-faced bear. "The skull is that of an old individual . . . and the teeth rather deeply worn," one report said.[7] The site—a "vast complex of oil, gas, and tar seeps" some 110 miles north of La Brea—didn't quite rival the sprawling goo pits in Los Angeles but provided important new context to life in the late Pleistocene, including that the two regions (one coastal, one desert) hosted a different suite of wildlife and that giant short-faced bears lived in both and probably had a taste for elk.[8]

In the summer of 1947, a field party from the University of Michigan's Museum of Paleontology found a partial skull of a short-faced bear in the rich Pleistocene deposits of Meade County, Kansas. At the time, they assigned it the name *Tremarctotherium simum*.[9]

Other bits of extinct bear turned up over the decades. More teeth, leg bones, and skull parts in Kansas. A clutch of toe bones, teeth, skull pieces, leg bones, and vertebrae in three Missouri caves. Skull fragments and leg bones in separate New Mexico caves. Toe and foot parts in eastern Oregon. Parts of the back half of a skeleton in Texas.[10]

Each find added another stroke of paint or two to the portrait of the vanished bear. Still, by the mid-1960s, almost ninety years after James Richardson emerged from Potter Creek Cave with the first skull, the story of the giant short-faced bear remained fractured, sketchy, and incomplete. There even seemed to be little consensus on how, exactly, to refer to it.

One of the inevitable first questions that arises upon the discovery of a new species is this: What in the world are we going to call this thing? And if you come up with a name, will others agree to call it that too?

The Swedish naturalist Carolus Linnaeus in the 1700s attempted to get us all on the same page by devising a brilliant, two-word system for describing all living things. The first word was the genus (a broad group) and then the species (what makes it different within the genus). Hence *Homo sapiens* for modern humans (from the Latin "wise man"), *Canis lupus* for wolves, and, one of my favorites, *Gulo gulo*, the scientific name for wolverines that describes a species so endlessly famished that simply using the Latin word for "glutton" once didn't suffice. But names are tricky things, especially when species are newly discovered and scientists aren't quite sure what they're dealing with. In the decades after the first giant short-faced bear was discovered in Potter Creek Cave, the scientific names were all over the place: *Arctotherium simum*, *Arctotherium yukonense*, *Tremarctotherium*, *Dinarctotherium*, *Arctodus*. Were these really all the same bear?

Finnish paleontologist Björn Kurtén attempted to cut through this taxonomic thicket in an obscure March 1967 paper that appeared in *Acta Zoologica Fennica*, an English-language journal published by the Finnish Zoological and Botanical Publishing Board. Kurtén had become something of a favorite for me, and his fingerprints are all over the story of giant short-faced bears. A paleontologist and novelist, he seemed to live on parallel tracks, hopping back and forth between being a serious scientist obsessed with precision measurements and sober speculation and a wild-eyed zealot of the Pleistocene determined to relay it in all its vivid glory to any lay person willing to give a few moments of attention. Better yet, he harbored his own fascination with bears of all kinds and perhaps he came by it honestly: *björn* means "bear." Kurtén died in 1988, and, time and again, I wished I could pick up the phone to shoot the breeze about short-faced bears and the lives they lived.

It took me months to track down his 1967 paper. It was only available in print and only a couple libraries in the United States had the compendium that included Kurtén's "Pleistocene Bears of

North America: 2. Genus Arctodus, Short-Faced Bears." I'd seen it referenced often in scientific papers but the actual copy of the paper remained totally elusive. For a few weeks, I was preoccupied with getting my hands on it, often suffering tiny bouts of despair that I might never read it myself. Finally, the library at the University of Wisconsin put me out of my misery with a kind loan to my local county library in Tucson. I had to laugh at the wave of excitement that washed over me when I picked it up—did my vision really blur for a few seconds?—knowing that its historical dissection of bear nomenclature would be 100 percent snooze-inducing for 99.9 percent of people on the planet.

Kurtén was a professor at the University of Helsinki in 1967 but spent long stretches in America, not only trying to get a fix on the short-faced bear but to understand the Pleistocene itself, especially the tumultuous lives of the mammals who'd crunched across the hard snow in the north and lolled around watering holes and tar pits in the south. He, too, had been following the trail of evidence around the great bear, pulling together studies and tantalizing scraps of information in the hopes of painting a clearer picture. Or, at the least, finally settling what to call the thing. "The short-faced bears have had a complicated taxonomic history," he said plainly at the top, as if clearing his throat before telling the tangled yarn.[11]

He started with Joseph Leidy's 1854 use of "Arctodus" when describing the tooth found near the Ashley River, the first scientific evidence of this kind of ancient bear in the United States (*arctos* being the Greek word for "bear," *odus* coming from the Greek word for "tooth"). That term fell into disuse after scientists years later found the origin of the tooth to be "indeterminate," meaning there was a question about whether the tooth may have simply belonged to a previously known bear, like a brown bear or black bear or perhaps those more known from South America.

Kurtén noted that most scientists until the mid-1920s followed the lead of the renowned E. D. Cope, who described the bear found by James Richardson in Potter Creek Cave in 1879 as *Arctotherium simum*. Cope had borrowed *Arctotherium* from Auguste

Bravard, a French mining engineer turned fossil hunter who used it in describing an ancient fossil of a giant bear in 1857 that had been found in Argentina's La Plata Basin, not far from Buenos Aires. *Simum* derived from the Greek word for "snub- or blunt-nosed." In 1926 a paleontologist from Buenos Aires named Lucas Kraglievich attempted to distinguish those giant bears found in North America by calling them *Tremarctotherium*. He, apparently, was unaware that another scientist, a man named Erwin Hinckley Barbour, had tried a similar trick a decade earlier when he used *Dinarctotherium* to describe the arm bone of a huge bear found in Nebraska in 1916.

The confusion led to a variety of names being applied to the great fossilized bears being found in the United States in the first decades of the twentieth century. In his paper Kurtén set out to set the record straight, including by poking a hole through the idea that the tooth Leidy described back in 1854 couldn't be conclusively connected to any particular species. "The second lower molar of the short-faced bears happens to be a particularly characteristic and easily identifiable tooth, and there is no doubt that the type belongs to this group." Based on that, Kurtén reasoned, the bears should rightly be referred to as *Arctodus*. That, after all, would follow a long-running principle in zoological nomenclature that the name first used in describing a new species should be given precedence unless there's a clear flaw with that approach.

The next question became how many types of short-faced bears should get their own unique name? They seemed to come in a wide variety of sizes, and some were so large that they seemed to warrant their own classification. Lawrence Lambe, for instance, suggested that the gargantuan specimen from the Yukon he described in 1911 ought to be called *Arctotherium yukonense*. Others suggested that the smaller versions also get their own distinguishing name.

Kurtén took a decisive, clear-eyed approach. While short-faced bears may come in many sizes, the structure, number, and placement of their teeth—so crucial in understanding and classifying the animal world—"is much the same throughout," he said. The size differences in the fossil record were likely more the result of sexual dimorphism between males and females and simply where the bears lived, what kind of food was available, how ample it was, and other factors that can put a strain on a body in the wild. (Kurtén wasn't the first to offer sexual dimorphism as an explanation for size differences, but his endorsement helped it to gain traction.)

His study of all of the available records at the time—dates, measurements, comparisons—led him to believe that it was justifiable to simply divide giant short-faced bear species in North America into two: the big, bruising *Arctodus simus* and its smaller relative, *Arctodus pristinus*, also known as the lesser short-faced bear. The latter wasn't uncommon, especially along the Atlantic Coast. More than 150 lesser short-faced bear specimens have been found, including in Florida, Maryland, and Pennsylvania, and that tooth from South Carolina described by Leidy in 1854. It was slender, with a lighter build and longer face, and the males were likely about the size of female giant short-faced bears.[12]

The lines drawn by Kurtén have proved durable in the ensuing decades with most short-faced bears relegated to either *Arctodus simus* or *Arctodus pristinus*. Not bad, especially given the historically tangled taxonomy of many bear species around the world.

And then, as if on cue, a few months after Kurtén's paper was published, one of the finest specimens of an *Arctodus simus* ever found turned up in an Indiana cornfield.

PART 3

Bones

13

What Happened in Fulton County

The bear had suffered and I felt bad. One day, roughly thirteen thousand calendar years ago, he had come to this place where I stood now in central Indiana, his front legs throbbing from sores that never seemed to heal. He was probably hungry—weren't bears always hungry?—and alone. The sun had come up over the Pleistocene's Middle America that morning. If it was fall or winter, it would've been muted light that slipped through the broken, mostly spruce forests of his home territory that stretched across hundreds of square miles. If it was summer, he may have lumbered slowly through chest-high grasses and paused to slurp out of one of the kettle-shaped lakes left behind by the retreating glaciers.

This place had provided for him, most importantly food. He needed flesh and anything else he could convert into calories to fuel a body that had swelled to more than 1,700 pounds. On his hind legs, he stood some ten feet tall—and taller if he reached his long arms toward the sky. When you're the biggest meat-eater in all the land, you have ample options for a meal. In his case, the menu included beavers, peccaries, deer, and even the occasional mammoth, mastodon, or giant sloth (not to mention a variety of plants). He couldn't afford to be too picky, though—by one estimate a giant short-faced bear needed to consume about thirty-five pounds of food a day.[1]

Although his claws were quite capable of tearing the hide off any animal and his jaws could crush any skull he could fit in his mouth, he didn't have to do all the killing himself. Chasing a pack of dire wolves off a freshly killed horse worked just as well. So did stumbling upon a newly rotting carcass of a peccary. Scavenging and even thievery carry no shame in the wild.

Still, it's never easy to be a bear, but things had become more difficult in his final days. Both of his forearms had wounds that had become infected. His body's immune system kept the infection from spreading, but the wounds seemed to be in a constant cycle of flaring up, draining, partially healing, and then starting over again. It had gone on for years and it hurt. In some wild animals, unhealed wounds can be fatal, especially when the pain saps their will or ability to eat. Once starvation takes hold it doesn't take long for the end to come. For this bear, for whatever reason, the wounds persisted and so did he, in a strange kind of throbbing purgatory.

He probably spent most of his time on this planet alone. Bears tend to stay close to their mothers during their first few years and then strike out on their own. Adult females soon have cubs. Adult males are perennial loners, ornery bachelors endlessly roaming, occasionally facing down another unfriendly bachelor, occasionally finding a female to mate with before traveling ever on. It's an isolated existence but one that works to keep the greater species going.

So here he'd come, to the edge of this lake. At some point the bear keeled over and breathed his last. And then the weather changed, the seasons rushed past and the decades and centuries wore on. Soon enough the Earth, and the efficient biological network that recycles us all, swallowed his entire body, ate away the flesh, and left only the bones.

For years there were reports of a monster in Fulton County. The stories said fierce and devilish creatures plied the gray-green waters of Lake Manitou on the outskirts of Rochester, a farming settlement in central Indiana. The manmade lake, created after a local river was dammed, was the work of the U.S. government in 1828 as part of a treaty with the Potawatomi Indians, who were going to operate a grist mill on its banks. Members of the tribe were forced to march to Kansas a decade later as part of the Potawatomi Trail of Death that killed more than forty people. The lake remained and the stories of monsters—usually in the form of serpentine visions rising above the surface—persisted too. Stay away from Devil's Lake, the warning went, lest you get too close to something so otherworldly.

In 1849 a shocked fisherman pulled a fish weighing "several hundred pounds" out of the lake, its blocky thirty-pound head put on display.[2] Forty years later, it took four men to haul a giant catfish out of the lake. It was laid out down the road near the county courthouse. A peek cost ten cents.[3]

Perhaps there were no monsters after all in Fulton County, just some grotesquely overfed fish. Or perhaps the real monster was yet to come.

More than a century later, in the fall of 1967, the Northern Indiana Public Service Company sent a small crew out to a farm a few miles south of Rochester to dig a routine trench and attach a valve on a gas line that was going in. The gas pipe traversed rolling corn fields and crossed over Mud Creek, a narrow band of slow-moving brown water. Per usual that time of year, the ground was wet and the muck was thick. A man named Amos Peterson was maneuvering his Caterpillar when the mud gave way and his tractor slid into the trench. A second man, Charlie Wargo, went to help, plunging the blade of his tractor into the earth to clear a path for Peterson's rig to be freed. He cut five or six feet into the ground and then work stopped because there, in the wall of dirt, was a scattering of half-buried bones, large and reddish-brown, touching the fresh air for the first time in thousands of years.

Peterson spotted the skull and pulled it free. The backhoe had split it, and the jaw was missing. Telling the story later, he said it was immediately clear he was holding the skull of some kind of bear. Grizzly bears were long gone from Indiana and black bears were rare. This wasn't either of those; it was way too big. The more they dug around in the brown slurry, the more bones they found, and they were huge.[4]

When it was later measured, the skull was more than 17 inches from the tip of the snout to the rear, and more than a foot wide measured from cheekbone to cheekbone. The other parts were impressive too: an upper arm bone almost 2 feet long, 21-inch forearms, a 25-inch thigh bone, an 18-inch shin bone, and shoulder blades nearly 19 inches across. A Fulton County monster, indeed.[5]

Work on the gas valve resumed while some of the men fanned out the bones atop the gas line to dry. Then what? The crew's foreman, William Lane, came over for a look and decided the bones needed the eye of a scientist. "He had enormous teeth on him, my goodness! I scooped them up and we argued about what it was—a cow, a sabertoothed tiger and what not. I had it all but just a few fragments. I got the rib cage, the thighs, the legs, the feet, the arms, the whole shooting match," Lane said later.[6]

The bones were left at the site, and Lane returned over the next several weekends, loading as many as he could into a wheelbarrow and then the trunk of his car. He also had the property owner, a farmer named Chet Williams, sign away any rights to the bear. Lane drove the bones an hour or so north to Westville, where Greta Woodard, a South African–born scientist, worked as an associate professor of biology at the north central campus of Purdue University. She then took the bones to the Field Museum in Chicago to compare them to other carnivoran specimens. In February 1968 photos of Woodard and Lane were on the front page of the *South Bend Tribune*. He in a suit, she in a white lab coat, they held the skeleton for the cameras. Woodard offered her tentative conclusion about the bones: "They are the fossilized remains of an immense bear."[7]

The importance of the discovery soon became clear: It wasn't just the first giant short-faced bear ever found in Indiana. It was the most complete skeleton ever found at that point.

14

Real Monsters

I flew from Tucson to Indianapolis on a red-eye in mid-spring, spent the rest of the night in a shabby hotel near the airport and got up early to make the two-hour drive north to Rochester. I'd never been to Indiana except maybe to change planes at some point, and a Hoosier friend provided an odd warning before I left: "There isn't much there, so don't expect too much."

He was wrong, as so many of us are about the places where we come from. Working my way north from Indianapolis on Highway 31 the next morning, the rolling land rose and fell easily under the wheels of the rental car like long, hypnotic ocean waves. It was green and fresh after a recent rain and the moisture clung to the air, cool and easy. Squinting just right, it wasn't hard to imagine a tall *Arctodus* lumbering out of a clump of roadside trees, nose high with purpose in the chilled morning.

I didn't bother with the town of Rochester at first. Instead I drove straight to where I'd read the bear had been found back in 1967. I'd written down the rough location, gleaned from magazine accounts and a little mapping I'd done before I left. Just a few miles west of Nyona Lake, not far from Highway 25, along Mud Creek where 600 South and 125 East meet up. Soon enough I arrived at an utterly ordinary spot surrounded by harvested cornfields, a few scattered farmhouses, and bushy clumps of trees that blocked the view to the farthest horizons on all sides. But where precisely had the bear been unearthed? The bear skeleton had been one of the most studied and famous at the time, but I knew from my reading there was no plaque or marker at the site. I'd have to rely on the locals to get me there.

From the road, I spotted an industrial-size garage with the bay door open. Three men inside were working on a piece of farm equipment. Swallowing my awkwardness from showing up out of the blue, I asked if they knew anything about the bear that had been found nearby more than fifty years ago. To my surprise the first man knew exactly what I was talking about and immediately pointed through a window to a brown patch about a quarter of a mile away. "There isn't much to be seen out there anymore," he shrugged. "It's just dirt."

A few minutes later I found another neighbor. She was out raking a spot where a tree and its stump had been pulled out a day earlier. She knew about the bear too and pointed me in the same general direction, suggesting I might return later in the day so she could share with me a couple newspaper clippings. (I did, and she came through.)

I knocked on the door of a third neighbor, this one closer to the spot the others had pointed me toward. I felt like a hunter slowly homing in on a target, albeit one that had left a long time ago. She invited me in, leading me through the living room where a man was watching the Chicago White Sox on TV. He didn't bother to turn around to see the stranger at the door, much less engage in my query about the bear skeleton. Through the picture windows on her back room, she finally gave me the definitive location I was looking for. Was there anyone left who was there the day the bones were found? Nope, they had all died, the neighbor said. "But we all know the story."

I pulled my rental car off the gravel road a mile or so away and tromped about five hundred yards through a cut cornfield. It was brown and woody and soft beneath my feet, speckled with unharvested ears of shriveled corn. A red-winged blackbird greeted me when I reached the edge of Mud Creek where the gas pipeline crossed, the same one they'd been working on in the fall of 1967. Perched atop a short yellow and white utility pole, the bird shrieked to me as if to say, "This is the place," and then took off in the roller-coaster way that they sometimes fly.

I stood in the same spot for a long time, suddenly somber as if I was visiting a place where a relative had died. The wind hissed in the knee-high grass along the banks of Mud Creek, and I tried to transport myself back thirteen thousand years. Instead of corn and soybean fields, there'd be a thick forest and easy meadows. Maybe the bear had been here on a spring day like I was and the clouds had opened up to let in some of the blue. Maybe even the red-winged blackbirds came to life as the temperature rose and the dew burned off and the insects became easier pickings. Maybe.

I stayed for close to an hour and could see occasional pickup trucks slowing on the road nearby when they caught site of me, a curious visitor standing alone in a desiccated cornfield. No doubt I was trespassing, but I had no idea against whom. My mind had gone somewhere else.

Just a few days earlier, the United Nations had issued a wrenching report that roughly one million animal and plant species were on the verge of extinction, "more than ever before in human history." Many could wink out within decades. The roster of the at-risk included 40 percent of frogs and other amphibians, about 30 percent of reef-making corals, and more than a third of all marine mammals, such as whales, dolphins, and sea lions. Estimating overall rates of extinction is tricky business—we still don't have a firm handle on how many species are on this planet, especially when you get down to the fungal and bacterial levels—but the UN report estimated that global extinctions are happening at "tens to hundreds of times" the average over the past ten million years. "And the rate is accelerating," the scientists added ominously. The report rightly made headlines across the world, one of those rare wildlife studies that seems to cut through the modern psychic clutter and put a scare into people, at least for a few hours until something else devoured it in the news cycle.[1]

The report's findings were still rattling around my head at Mud Creek, especially the flimsy hold on existence that so many of us have on the planet. The giant short-faced bear rode a particularly potent wave of extinction in the late Pleistocene. Over the course of a few thousand years, about half of the large mammals in North America had been wiped out of existence. Mammoths, mastodons, camels, giant sloths, saber-toothed cats, lions, dire wolves . . . they were all just gone, swept off life's stage in a heartbeat. And *Arctodus* went with them.

Time has always been short for most species on this planet. Heck, for the better part of several hundred million years there really *was* no life here. Earth was a soupy, poisonous, volatile place hostile to anything as audacious and optimistic as a living thing. Bacteria came first around four billion years ago, and things have never been the same since. For all the marvels of every plant, animal, fungi, and other forms finding their way into existence, there has always been a natural cost. The world changes, continents shift, weather patterns spin anew, prey find new homes, predators follow along in response. Some species weather the storms for a long time by evolving, adapting, or just being damn tough. Others are simply overtaken by events and slip into extinction. Mammal species, for instance, tend to last roughly a million years or so on the planet (though some much longer and some less so).[2]

Scientists estimate giant short-faced bears existed from around two million years ago until around eleven thousand years ago. Not bad for a big beast that requires a lot of energy to survive. Certainly by the time the Rochester bear died, the storm clouds of extinction for his particular species were already gathered and swirling overhead. He wasn't the last, but he was among the last— certainly many more giant short-faced bears had come before him than would come after.[3]

After a long while in the cornfield, I raised an imaginary toast to him and started back to the car. From the side of the road, I looked around to make sure I wasn't missing a sign or a little monument designating that this was where the bear had been found. There was only a stubby utility pole marking the gas line that had been the focus on the work back in 1967: "Before digging, please call."

After the cornfield, I drove into Rochester. It's a quiet farming community of six thousand with a Romanesque limestone county courthouse downtown, a shuttered movie theater, empty store-fronts, and a few local shops clinging to survival in the wake of Walmart's arrival on the outskirts of town. "Do you know any-thing about the bear found outside of town?" I asked an older man outside of a medical supply store near the courthouse. He gave me a funny look and then smiled pleasantly. "I'm afraid I do not."

I walked through town looking for any sign that might hint at the historic *Arctodus*. Every small town needs a bit of notoriety, something to call its own, even a corny way to stand out from all the other small towns in America. What could be better than a giant short-faced bear? Momentarily hopeful, I took a minute to check and found that the mascot for the local high school was . . . the Zebras. *Come on.*

I thought about New Bern, North Carolina, a city of about twenty thousand along the Neuse River. It's lovably and fully obsessed with its bear heritage, namely its namesake city in Switzerland (*Bern* roughly means "bear"), which has its own storied history with bears going back to its founding in 1191 at the place where several brown bears had once been hunted. New Bern's logo fea-tures a bear silhouette with a red curling tongue, and there are cast-iron bear heads mounted on city hall and the fire station. There's even a special license plate featuring the bear on a red and yellow crest. I first heard about it after Hurricane Florence swept through in 2018 and toppled some of the fifty fiberglass bear statues—each six feet tall on its hind legs and colorfully painted by a local artist—erected around town for its three-hundredth anniversary. There was a memorable news photo that showed the broken body of one of the bears in the middle of flooded streets and, in the background, another one of the statues still standing erect, decorated in a colonial-era white wig and red coat. "Some of our beloved bears have wandered away," the city said wistfully in a tweet after the storm.[4]

No sign of a beloved bear here in Rochester. No *Arctodus* statue on the county courthouse lawn. No banner above Main Street, no special license plate.

"Why is that?" I asked Shirley Willard, Fulton County's historian, later in the day.

"I don't really know," she said, pausing as if considering the question for the first time. "But someone should do that! Have a parade or something."

Willard was one of the few people who had written about the Rochester bear, a handful of stories over the decades that appeared in local newspapers and magazines. The idea of a such a massive bear in mostly bearless Indiana farmland fascinated her from the start, and she could never understand why others didn't share her interest. I met her at the Fulton County Historical Society, a long white rectangular building just off the highway about four miles north of Rochester. She took me to the back room that had the rumpled, lived-in look that every county historical society office seems to have. Books, papers, and boxes of photos and memorabilia covered tables and desks, and historic photos were up along the wood-paneled walls. Willard, warm and businesslike and a little quirky, as a small-town historian should be, took me straight to a blue filing cabinet in one corner and opened up one of the drawers. "There it is. That's where we keep the giant short-faced bear," she smiled, and then waited for me to turn to her so she could raise her hands in big claws. "Roar!"

Her records were mostly the stories she had written about the Rochester bear and a few memories she had collected when she spoke to some of the work crew who were at the site in 1967. We chatted for a long time around a wood table in the back, and the conversation turned to the monster in Lake Manitou. Willard said she still gets a story every year or so from someone who said they saw the big snake-like fish along the shore or beneath their boat. "Of course no one wants their name used, but you should hear them," she said.

And the Rochester bear? Did it register locally beyond the county historian?

"I could write a story about it right now and people here would say, 'I've never heard of it.' It's just not very well known," Willard said and then laughed: "In fact I think people know more about the serpent than the bear."

15

A Hoosier's Search

After making front-page news in Indiana in early 1968, the bones of the Rochester bear went with Woodard at Purdue's north central campus. She teamed up with William Turnbull of the Chicago Field Museum and hatched a plan to do an in-depth study on the bones—to figure out everything they could about this beast from Rochester and what it might tell scientists about life in the Pleistocene. The plan was short-lived. Woodard died unexpectedly, and her husband brought the bones to the Field Museum in 1970, where they were put into the fossil mammal collection. Although they were once used as a model for a painting of the bear, the bones generated little interest until a curious scientist named Ron Richards showed up a decade later.

Richards, a Hoosier through and through, grew up in Indianapolis and started collecting fossils when he was ten. When he was a high schooler in 1964, the Boy Scouts took him to his first cave, and that led to his first find: the jaw of a black bear. "I was hooked," said Richards, who is now the senior research curator of paleobiology at the Indiana State Museum in downtown Indianapolis. As a young man, he wore out his copy of *Mammals of Indiana*—chock-full of photos of teeth and skulls—and soon expanded his obsession to include all manner of fossils, from ancient snails and mussels to shrews, mice, and eventually mastodons and mammoths. Indiana, it turned out, was a hotbed for fossils, from the harboring darkness of the caves in the southern part of the state to the expanses of soggy marl and peat in the north, perfect for preserving bones across the eons.

In 1980, early in his career as a paleontologist, Richards was surveying regional museums in search of Pleistocene mammals from Indiana. At the Field Museum in Chicago, walking down a long corridor packed with storage cabinets, "my eyes locked on the label of one of the cases: 'Arctodus.'" He was shocked by the stunning array of massive bones inside.[1]

"My first thought was, 'Wow, this is from Indiana? This is important,'" Richards told me. "I was used to dealing with black bears, but that was hardly a bear compared to Arctodus. This is not a frail little animal. It's just such a monster."

Turnbull, the Field Museum's curator of fossil mammals at the time, was with Richards that day. Both understood the importance of the Rochester bear and the need for someone to conduct a detailed scientific description. Turnbull was busy with national and international paleontological work but agreed to provide some guidance if Richards would do the bulk of the labor on the bear, which promised to be substantial. Describing the nearly complete skeleton for a scientific publication would mean measuring— down to the last millimeter and from every conceivable angle— each bit of the bear's bones. That included more than twenty-five teeth, thirteen vertebra, seventeen ribs, plus arm bones, leg bones, the pelvis, shoulder blades, and the skull.

Because of the bear's rarity, the Field Museum would loan out only portions of the skeleton at a time to Richards. That meant borrowing a car from the Indiana State Museum, or from the state motor pool, and making the three-hour drive to Chicago several times. It was a meticulous process, done often on weekends, to measure and notate each piece, describe it and compare it to other specimens of Arctodus simus described in the scientific literature over the previous century. "It took forever," Richards said. "But bit by bit we got it done."

Along the way, as Richards tried to describe the bear and put it in a larger context, it became clear no one had compiled a comprehensive history of all the giant short-faced bears ever found in North America. A few other scientists had stopped and started, but it'd never been completed. The details of the short-faced bears' existence were scattered willy-nilly across scientific papers, letters, notebooks, museum collections, and idle chit-chat among scientists that may or may not have pertained to the actual species. It was no secret that parts of these ancient bears had been pulled from the goo of La Brea in Los Angeles, removed from a historic cave in North California, and plucked from the dark soils of the Yukon after discovery by gold diggers. Some of them had been extensively studied and breathlessly reported. Other discoveries were known by just a few people. But no one could say with any certainty how many specimens had yet been found. Dozens? Hundreds? More? And how did they compare with each other in size and age?

This would be the next task. For the next decade Richards and Turnbull scoured scientific literature for references to the bear. They visited museums and talked to as many scientists as they could. They chased rumors, dismissed dead-end leads, exchanged letters with curators, and spent hours on the phone. It was a mess. In some cases, the bears were misnamed, or *Arctodus simus* was confused with its smaller cousin, *Arctodus pristinus*, or even listed under outdated names—like *Arctotherium*—that Kurtén had tried to do away with in his 1967 paper. Sometimes there were dates and locations; sometimes there was little more than a cryptic reference to a fossilized, bear-like bone. Grizzly bear bones were mixed in with giant short-faced bear bones. Some museums didn't bother to publicly state they had *Arctodus* specimens in their collections.

For help, Turnbull directed Richards toward Clayton Ray at the National Museum of Natural History in Washington DC. Ray, a fellow Hoosier and intrepid Pleistocene researcher, had launched his own project compiling information about giant short-faced bears. Ray eventually realized his busy schedule would probably never allow him to finish the project, so he handed his data over to Richards. He also urged him to check out the *Arctodus* collection in the drawers at the American Museum of Natural History in New York. That's how Richards ended up in Manhattan in February 1990 in a hotel right across the street from one of the most important natural history museums in the country. It was a wonderful sort of trip, if a bit feverish.

"I'd spend all day there and then I'd go back at night and work till 11 or 12," Richards said. "A lot of times I was the only one there. That's how I like to work."

There were endless corridors with cabinets, drawers, and boxes with all manner of fossilized animals inside, short-faced bears and grizzlies, of course, but also other species like bison. It was hard not to wander into the other rooms and see the works of the great paleontologists like Henry Fairfield Osborn, one of the most famous dinosaur hunters in the late 1800s and early 1900s, who spent twenty-five years as president of the museum. "Here I was walking where they'd been and thinking, 'Maybe I *am* a real paleontologist.' It was kind of a self-realization," Richards said.

Meanwhile, unbeknownst to Richards and Turnbull, a University of Toronto scientist named Rufus Churcher had also been trying to get his arms around the full picture of short-faced bear fossils. They eventually stumbled into each other and merged their efforts. All told, after years of research, they revealed there were more than one hundred *Arctodus simus* specimens distributed across more than thirty museums in Canada, the United States, and Mexico.

In 1995 Richards and Turnbull had a thirty-four-page paper published in the Field Museum's journal *Fieldiana* that described the Rochester bear in detail. It had detailed photos and measurements of every single piece, from the top skull and every tooth to the ribs, legs, and pelvis. It made comparisons to other giant short-faced bears. It discussed pollen found around Mud Creek, noted that its skeleton had likely been deposited in a shallow lake near a forest dominated by spruce trees, and speculated that the bear's local menu likely included giant beavers, mammoths, mastodons, moose, deer, ancient bison, and musk ox. He'd stood about five-foot-two at the shoulder, and radiocarbon dating estimated he lived around thirteen thousand calendar years ago.[2]

The following year, Richards, Churcher, and Turnbull published "Distribution and Size Variation in North American Short-Faced Bears, *Arctodus simus*." The twenty-two-page paper came out in a University of Toronto publication commemorating Churcher's retirement. There are only a few dozen hard copies scattered around the country today, but the paper has become a touchstone for anyone studying giant short-faced bears. Especially valuable is the exhaustive appendix discussing every single specimen down to its location, date of discovery, and painstaking details such as whether a tooth was a left molar or right, if the skull had a fracture or if the left thigh bone was complete.[3]

When I talked with Richards more than two decades after the study was published, I expressed my admiration and thanks for such a heroic effort. The bear, often shunted to the side of Pleistocene lore, had finally gotten the kind of intensive review it deserved, I told him. He demurred. "Oh, it was fascinating to work on," he said, adding with a laugh, "but I'd never do anything like that again."

After visiting Rochester and the site where the bear had been found, I woke up early the next morning and drove to Chicago to see the bear for myself. The Field Museum cuts an imposing figure on the shore of Lake Michigan with it its neoclassical frame built from more than three hundred thousand cubic feet of Georgian white marble. As I parked, I noted that the Rochester bear's final resting place, on the third floor of the Field Museum, is right across the street from another modern ursine palace of sorts: Soldier Field, home to the NFL's Chicago Bears. Bears are everywhere.

When I bought my ticket, the agent at the counter made a point to mention where Sue, the forty-foot *T. rex*, could be found: third floor, in the *Evolving Planet* exhibit tracking 4.5 billion years of life on Earth. "You can't miss it," he said.

"And you have a giant short-faced bear on display?" I asked. He paused to think. "Yep, I think there's one up there too."

He was right; you couldn't miss Sue. She has her own room in the exhibit, complete with an audio presentation and coordinating colorful light show. I don't begrudge Sue the attention—it's an extraordinary display that sends the imagination crackling, especially when you think about how quickly it could catch and consume any one of us who happened to be caught in the wrong place and epoch. But my quarry was still to come. I finally found him toward the end of a Pleistocene exhibit in the middle of a horseshoe-shaped walkway that also included posed skeletons of cave bears, giant sloths, bison, mammoths, mastodons, and a towering Irish elk with antlers that looked big enough to block out the sun. The Field Museum had speakers piping in icy wind gusts to enhance the Pleistocene mood, and the walls were lined with Charles R. Night murals, big colorful scenes of mammoths, saber-toothed cats, and a European cave bear.

I sat on the bench in front of the Rochester bear skeleton for the better part of an hour. He stood tall on his hind legs with both arms extended, huge fingers splayed in search of some phantom prey, his mouth wide open like he'd been caught in the middle of a roar that would never be completed. What percentage of his real life had he spent in that fearsome position? No doubt most of it was on all fours patrolling those forests and open grasslands, trying to find some food and get through his day. Still, it was impressive to seem him towering on both legs, maybe ten or eleven feet above me. Sue, a couple rooms over, might have had her way with him, but he would've given her a run for her money.

I felt a strange affinity for him, especially having been at his death bed less than twenty-four hours earlier along the shores of Mud Creek. It was a Wednesday afternoon in the spring, so the exhibit was quiet. Often it was just me in the room and I admit to taking the opportunity to chat with him, if only to extend my compliments on his stature and what I assumed were his ferocious hunting skills. And to ask how he'd gotten those wounds that never healed on his forearms. A group of teenagers on a field trip eventually rescued the poor bear from my monologue. They were holding papers with an assignment to learn about certain things. The giant short-faced bear was not on the list, and they kept moving. I wanted to stop them. "That's one of the biggest bears that ever lived! And it was found just a few hours from here!" But I held my tongue and let the injustice continue. After another period of quietude, another group came in, most of them fully inhabiting the awkward posture of middle schoolers. They'd understand, I predicted. Several stopped in front of the bear. "Whoa! No one's messin' with him," one of the young, antsy boys said. *Damn right*, I thought.

The bear and I communed for a while longer, and as I got up to leave, I remembered Shirley Willard's bear pose in the back room of the historical society in Rochester. I raised my hands at him the same way, fingers spread and ready to strike. "See ya next time, friend. Roar!"

I was supposed to see Ron Richards the next day at the Indiana State Museum in Indianapolis. We had spoken on the phone, but I wanted to meet him in person and check out the plaster cast of the Rochester bear they had in their collection. A colleague named Damon Lowe called me the night before to say Richards was ill, nothing terribly serious, but that he wouldn't be able to see me the next morning. Gamely, Lowe offered to show me around.

Lowe's a biologist and a senior curator at the museum, a sprawling and sparkling three-story building along the White River. When I showed up, he had just overseen the installation of a life-size cardboard replica of the first winner of the Indy 500 in 1911, a car called the Marmon Wasp. He'd worked with Richards for a long time and had been on his share of digs for Pleistocene fauna in Indiana. He was also well familiar with the *Arctodus* specimen from Rochester. "My fantasy football team was called the Short-Faced Bears," he smiled as he led me to one of the museum's back rooms. "I won."

The staff-only collections room he took me to was a wonderland of Indiana finds pulled out of the ground over the years. Teeth, skulls, and bones of dire wolves, peccaries, beavers, elk, mammoths, and mastodons all waited for us under the room's bright lights. Some bones were clean and light colored, offering hardly any hint they'd been pulled from the ground. Others were a chocolatey brown, the product of spending centuries in a darker kind of dirt. Deeper into the room, Lowe rolled back a towering vertical shelving system on wheels and pulled out a low drawer marked with a small card that read "*Arctodus simus*." Inside, like an artistic display, dozens of white cast bones fanned out across the foam-lined bottom: giant ribs arranged like opening and closing quote marks, vertebrae along the top, leg and arm bones on each side, both sides of the lower jaw next to each other like mirror images and a magnificent cast of the skull on the far right. The whole thing made me catch my breath.

On the one hand, I was taken aback by the size of the skull and thought about what the people on the work crew along Mud Creek must've thought when they pulled it out of the marl back in 1967. No wonder they knew pretty quickly it wasn't the remains of an ordinary black bear. But I was also struck, and a little saddened, to see this entire wild and fierce bear relegated to the strict rectangular confines of a museum drawer. I knew this was just a series of plaster casts and not the real bones, of course, but it left a weird taste in my mouth all the same. This bear had once been the king of all the land he'd laid his eyes on. In death, we'd found a way to tame and contain him in a sterile, climate-controlled storage room with nary an ounce of daylight or a whiff of fresh air to be found.

I didn't discuss any of this with Lowe. He stood back quietly and let me examine each piece and photograph what I could. He asked if I wanted to measure any of the parts, assuming I'd come on a scientific mission rather than an admiring, journalistic one. When I declined, he seemed content to just appreciate the bear with me. "Amazing, huh?"

As we parted ways, Lowe left me in the museum proper to wander, pointing me toward the Ice Age Graveyard display. It's an impressive site, especially the replica of a mastodon buried shoulder deep in brown muck and surrounded by trees and other Pleistocene animals. Nearby there was a wall-sized map showing how many mammoth and mastodon sites had been found in Indiana. Dozens and dozens are scattered around the state. As ever, it had me wondering how many animal remains are strewn beneath our feet, waiting only to be revealed by an errant scraper and shovel. That seemed especially true in Indiana. "Folklore relates that any bog larger than an acre likely has mastodon remains in it," Ron Richards had once quipped.[4]

Farther down the hall next to the giant sloth, I found the Rochester bear, or at least another plaster version of his skeleton. This time he was mounted in a hunched-over position, back feet firmly on the ground, left front paw on a rock and the right floating free. His neck was craning toward the ground and his mouth agape, like he was ready to pounce on a small animal. On the railing in front, there was a mounted replica of his huge lower jaw, brown and dangerous looking, that'd been well-worn by thousands of people who had rubbed it over the years, especially the thick bottom canine. Without thinking, I did it too. Even in this detached digital world, we still depend on our primal sense of touch to assess things from the natural world—teeth, shells, animal hides, tree bark. Maybe we're not as far from our ancestors as we sometimes think.

I found a bench but didn't linger long. I realized I'd seen the Rochester bear in three iterations over the past two days. Once at the Field Museum in Chicago where he'd stood tall, here in the Indiana museum display where he was stooped over and ready for action, and finally in the drawer that Damon Lowe had shown me. After thousands of years of peace and quiet buried in the peat and marl at Mud Creek near Rochester, he'd been unceremoniously excised, transported, examined, cleaned, stored, photographed, copied, measured, and put out for all to see. Everlasting life after death.

I left Rochester, south on Highway 31 in the nonexistent morning rush-hour traffic, back through the undulating farmland, past the barns and the idle machinery and the aging pickups parked on the grass. These places once existed in the shadows of enormous ice sheets, or beneath them, and gave a home to an array of extraordinary beasts that are gone now, left only to fight for footing in our own crowded memories. As I passed the sign for Nyona Lake and the road that had taken me to the bear's death site a couple days before, a wistful little wave washed over me, followed by a pulse of delight that he'd been here at all.

16

Dispossessed

A light rain fell a few weeks later as I pulled into the student parking lot at Sierra College, northeast of Sacramento. I'd come to see Dick Hilton and some of the bear specimens at the college's natural history museum.

Standing in the doorway and shaking the wet off my coat, another world hit me. Or, more precisely, several did. The skeleton of a sea cow hung overhead and dead ahead was a you-can't-miss-it glassed-in display of taxidermied bears, including a towering polar bear on its hind legs, an Alaska brown bear standing next to him, plus a grizzly and black bear. In the lower righthand corner was the replica of a giant short-faced bear skull, looking dark and brooding. Elsewhere, the museum had mounted heads of all manner of beasts from around the world plus the remains of an ancient tortoise and a replica of a massive ichthyosaur that looked like it belonged on another planet. It was a bestiary par excellence that small herds of students hustled past, many with barely a glimpse.

I found Hilton in his cramped office off one of the main hallways. Russell Shapiro, the California State University—Chico paleontologist who accompanied me to Potter Creek Cave, had pointed me toward Hilton as a deep well of knowledge about California in the Pleistocene. Hilton, wiry and direct, was gracious to me, providing access to some of the college's vast files and, at one point, standing with me at the copy machine in a hallway alcove while I ran off page after page of historic documents. It's where I picked up a copy of a 1925 paper from John C. Merriam and Chester Stock—two early luminaries in the study of California's Pleistocene scene—that included a pencil sketch of what the head of a giant short-faced bear might've looked like. The sketch, under the supervision of Merriam and Stock, was done by Charles R. Knight, a well-known paleo-artist at the time. I've never seen the drawing anywhere else except for that paper and I've spent hours staring at it. It's a profile view of the bear's furry head with its ears perked up, eyes alert and looking forward and, beneath a short muzzle, mouth slightly ajar to offer a peek at a couple of large teeth. Knight gave the bear a gentle aspect. There's something human in the way it's looking off the page, like it's a happy spectator of an event we cannot see.

Later Hilton took me to a locked storage building near the museum. Inside a black cabinet, he pulled out a tray of giant short-faced bear bones, each in a little box, that had been found years before to the north, in Tehama County.

We chatted while I carefully looked over the specimens, not really sure what I was looking for but wanting nonetheless to examine each vertebrae, toe bone, and fragment. Hilton figured that the Pleistocene climate in Northern California was colder than it is now, but what really made it different was the number of gigantic species walking around: bears, sloths, cats, horses, and camels along with smaller species like deer, peccary, rodents, rabbits, turtles, birds, fish, and mussels. "It was incredible," he said. "It was like Africa." We marveled together at the size of the teeth in the trays in front of us. No doubt *Arctodus* held a top spot among the animals. "Nobody fussed with a short-faced bear unless it was sick or injured," he said. "It was like a king."

Toward the end of my visit, Hilton asked me a detailed anatomical question about the bear. I had no idea about the answer, and a flash of confusion went across his face that seemed to ask: *This guy's a scientist and he doesn't know this?* I felt bad and reminded him that I was not a scientist but merely a writer, an interested layperson, in pursuit of this bear that used to live among us but was gone now. Hilton was unfailingly nice for the rest of my visit, but at the end I couldn't help feeling, warranted or not, that he was left wondering: *What's this all about?*

I'd begun to wonder the same thing. It's true that bears have always held a bit of a spell over me, but this had become more than a passing interest. The obsession that slowly overtook me had me periodically stepping out of my day job so I could travel the country to be in the places where this bear had once been and touch the remains that had been left behind. There were dingy hotels, greasy gas station lunches, endless drives in dented rental cars, and wild goose chases into corners of the country I'd never thought to see. Classic rabbit hole. I was now combing century-old documents, searching for obscure scientific journals, and sitting on the tile floors of barely trafficked corridors of university libraries with books spread in front of me, some of them so stiff it felt like they hadn't been cracked open in decades. Boxes gathered in my house and shelves filled up with all manner of bear books. I tracked leads, pestered and phoned and wrote to experts, and blathered to friends and family about picayune paleontological details that almost certainly were interesting only to me (oozing arm sores of an extinct bear, anyone?). I'd become a bit of a stranger to my regular life and spent much of my time, at least in my head, dwelling on ancient bears lost and found. They all existed in a kind of foggy dreamworld that had become irresistible to my wandering mind during waking hours and asleep.

Over the months the venture morphed into something more: meditations on animals of grotesque size and supernatural strength, on the nature of predatory species, and eventually on the unalterable and mysterious finality of extinction, the permanence of death. At its core, perhaps, was an attempt to look behind the great curtain in the hopes of a glimpse to the other side.

I was ticking off all of these things over a corner shop burrito a couple of hours after my visit with Dick Hilton at Sierra College. After a long breath, I wrote two words on a scrap of paper. *What gives?*

My dad died not long before I started really fixating on giant short-faced bears. I thought about him a lot on my trips to these places where the bears had been.

He loomed large, as fathers will do: decent, funny, stubborn, sometimes cranky, proud of his two decades in the Navy Seabees, principled, and sporting an enviable knack for remembering people's names and stories. He was tough on his three sons sometimes, insistent that we learn how to change a flat tire, work a power saw, and understand the value of grueling, labor-intensive jobs as young men. The kind of things that build a work ethic. He was big on manners and firm handshakes and playing sports the right way. He taught me to fish and read a map and tell funny, self-deprecating stories. How to handle a clutch and decipher baseball's infield fly rule and build a gate for a backyard fence. And the important difference between a Gibson martini and the rest. He had a temper but worked on keeping it in check. I certainly wore his patience in my younger days and vice versa. We didn't agree on everything but grew closer once I moved deeper into adulthood, and he mellowed (only a little) in his later years.

We never really talked about bears, but I think the two became interconnected for me along the way, sometime after he died and I went off looking for bears that were no longer there. Extinction of a species and the death of a parent leave behind similar wakes. The world was a certain way when my dad was here, and it's a different place now that he's gone. It's the same with the bears. We keep living and the planet keeps spinning and most beings aren't cognizant of the loss, or that these bears even existed in the first place. For whatever reason, that had a peculiar sting for me. These beasts, so mighty and formidable and fascinating, could leave their mark on their world for more than a million years and then vanish and none of us, really, are the wiser. Are things really that transitory? And do those who disappear remain totally out of reach from the rest of us? I sometimes wondered if I could get closer to my dad by paddling Washington's Hood Canal, where his ashes had been scattered, and, in the same vein, whether I could somehow commune with this extinct bear by handling dispossessed bones from a museum drawer or crouching in a cave and inhaling the same stale air.

Of course, the intensity of the feeling is different. Losing my dad is a sharp and painful loss that will never find "closure," whatever that is. The disappearance of these bears produced its own sort of grief, a dull pang of mourning for an extraordinary species I wish was still lumbering through the wild. Both left me smarting that such potent forces could simply be snatched from us before we're ever ready. The world's an unfair place and existence is a fleeting, quicksilver thing.

My dad probably wouldn't have wanted to see this bear journey become so sentimental and complicated. Puzzling out the nature of death and disappearance wasn't something we'd get into. Too private and abstract and unknown. Instead, always curious, he would have asked me questions about what I'd seen, who I'd talked to, how these bears worked, the logistics of the arrangements, and where I'd been. *Can you show it to me on a map?* He would've laughed that I was doing this at all and toasted me with a beer in a frozen mug when it was all through.

Whatever the case, the two losses occupied the same places inside of me and exposed the kind of universal questions that we all eventually wrestle with, the ones that ask about the nature of existence, the fundamentals of our being and the histories we inherit, create, and later pass to those who come after. I left Sacramento the next day, unsettled with the knowledge that I was searching for answers to questions I couldn't quite formulate but determined to do it anyway.

Fig. 1. (*above*) Artist George Teichmann's rendition of a North American giant short-faced bear, known also as *Arctodus simus*. Parts of more than one hundred giant short-faced bears, which lived during the Pleistocene, have been found so far, from Alaska and the Yukon to California and Oregon, across the Midwest and even Florida. Giant Short-Faced Bear © Artist George "Rinaldino" Teichmann, 1999.

Fig. 2. (*below*) Teichmann's comparison between giant short-faced bears and other species, including grizzlies, brown bear, and black bears. The biggest among the North American giant short-faced bears, the males, weighed close to a ton and, standing on hind legs, towered more than ten feet tall. North American Bear Comparison © Government of Yukon / Artist George "Rinaldino" Teichmann, 2002.

Fig. 3. The entrance to Potter Creek Cave in Northern California. The first skull of a giant short-faced bear was found in the cave in 1878 by a man named James Richardson. Parts of several more giant short-faced bears were eventually recovered from the cave, along with remains of more than twenty other extinct species. Photo by author.

GEOLOGY AND PALÆONTOLOGY.

THE CAVE BEAR OF CALIFORNIA.—In exploring a cavern in the Carboniferous limestone of Shasta county, Cal., James D. Richardson discovered the skull of a bear beneath several inches of cave earth and stalagmite. The specimen is in a good state of preservation, and demonstrates that the cave bear of that region was a species distinct alike from the cave bear of the East (*Ursus pristinus*), and from any of the existing species. In dimensions the skull equals that of the grizzly bear, but it is very differently proportioned. The muzzle is much shorter, and is wide, and descends obliquely downwards from the very convex frontal region. It wants the large postorbital processes of the grizzly, but has the tuberosities of the polar bear (*U. maritimus*), which it also resembles in the convexity of the front. Sagittal crest well developed. Three (one median and posterior) incisive foramina: three external infraorbital foramina. The teeth are large, and the series presents the peculiarity of being without diastema. The crowns of the premolars are not preserved, but if there were not three premolars, the second tooth has two well developed roots. First true molar with but two external and one internal tubercle. The absence of diastema renders it necessary to separate this bear from the true *Ursi*, and I propose to regard it, provisionally, as a species of *Arctotherium* Gerv. The canine teeth are large and compressed at the base. Length of cranium along base from below apex of union to premaxillary border, m. 0.387; length to posterior nares, .202; elevation of forehead vertically above the posterior extremity of the last molar, .141; width between inner border of posterior molars, .076. The species may be called *Arctotherium simum.*—*E. D. Cope.*

Fig. 4. Edward Drinker Cope's brief, 285-word announcement of the find at Potter Creek Cave appeared in December 1879 in *The American Naturalist*. He called it "the cave bear of California" and named it *Arctotherium simum*, though it was later renamed *Arctodus simus. The American Naturalist* (1879), University of Chicago Press Journals.

PLATE XXI.

Arctotherium simum Cope; ⅓.

Fig. 5. These drawings accompanied Cope's second paper on the Potter Creek Cave bear skull, published in 1891 in *The American Naturalist*. "To judge by the skull alone," Cope wrote, "the Californian cave bear was the most powerful carnivorous animal which has ever lived on our continent." *The American Naturalist* (1891), University of Chicago Press Journals.

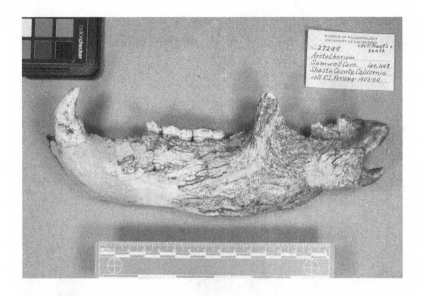

Fig. 6. The lower jaw of a giant short-faced bear found in Samwell Cave, California, in the early 1900s. University of California Museum of Paleontology, 2014.

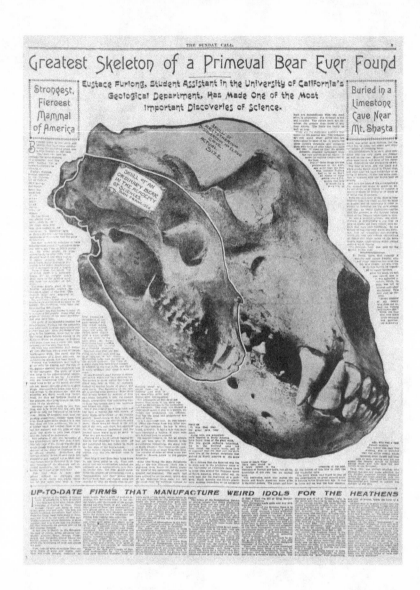

Fig. 7. The *San Francisco Call*'s full-page spread on September 14, 1902, detailed the discovery of more giant short-faced bear bones in Shasta County. Courtesy of California Digital Newspaper Collection, Center for Bibliographic Studies and Research, University of California–Riverside, http://cdnc.ucr.edu.

Fig. 8. Des Moines University paleontologist Julie Meachen descends into Wyoming's Natural Trap Cave. Thousands of bones have been found on the cave floor, including the partial remains of two giant short-faced bears. Photo by Justin Sipla.

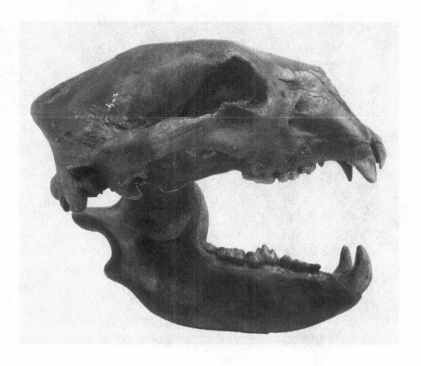

Fig. 9. (*above*) A skull cast of a giant short-faced bear found in the La Brea Tar Pits in Los Angeles. Courtesy of the Division of Vertebrate Paleontology, Peabody Museum of Natural History, Yale University, peabody.yale.edu (*Arctodus simus* skull, YPM VP 55748).

Fig. 10. (*top right*) James W. Lytle, assistant vertebrate paleontologist, stands in the "bone room" in the basement of Los Angeles County's Natural History Museum with material recovered during extensive excavations at the La Brea Tar Pits between 1913 and 1915. Parts of more than thirty giant short-faced bears have been found at the site. Courtesy of the Natural History Museum of Los Angeles County.

Fig. 11. (*bottom right*) A diagram, produced as part of a history of the La Brea Tar Pits, compares the giant short-faced bear (*back*) with the extinct California grizzly bear. Courtesy of the Natural History Museum of Los Angeles County.

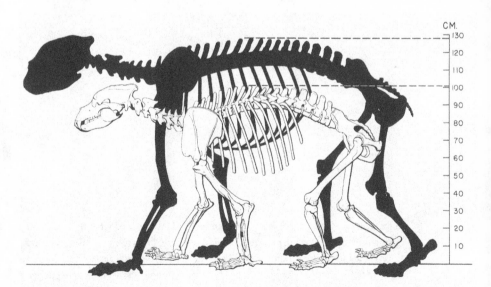

Fig. 12. In 1925 famed paleo-artist Charles R. Knight sketched what he thought the giant short-faced bear may have looked like. John C. Merriam and Chester Stock, "Relationships and Structure of the Short-Faced Bear, Arctotherium, from the Pleistocene of California," publication no. 347, pp. 1–35, courtesy of the Carnegie Institution of Washington, 1925.

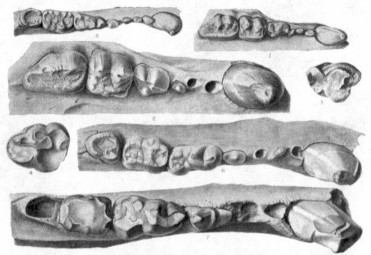

1. *Tremarctos ornatus* (Cuvier). Superior dentition, No. 171011, U. S. Nat. Mus. Recent, South America.
2. *Arctotherium californicum* Merriam. Superior dentition, No. Z 1, L. A. M. C. Pleistocene, California.
3 and 4. *Indarctos? oregonensis* Merriam, Stock, and Moody. P4 and M2, No. 22362, U. C. C. Pliocene, Oregon.
5. *Tremarctos ornatus* (Cuvier). Inferior dentition, No. 171011, U. S. Nat. Mus. Recent, South America.
6. *Arctotherium simum* Cope. Inferior dentition, No. 3001, U. C. C. Pleistocene, California.
7. *Agriotherium (Hyaenarctos) schneideri* Sellards. Inferior dentition, No. 6856, Fla. Surv. Coll. Pliocene, Florida.
All figures are occlusal views, ×0.873 (approximate).

Fig. 13. A 1925 comparison of the jaws and teeth of the giant short-faced bear and others. The short-faced bears' jaws are the longer samples on the second and third rows. At the top are samples from South America's spectacled bears, which are still extant. The others are from extinct bears from the Pliocene epoch, which preceded the Pleistocene. Courtesy of the Carnegie Institution for Science, from "Relationships and Structure of the Short-Faced Bear, Arctotherium, From the Pleistocene of California," by John C. Merriam and Chester Stock, from publication No. 347, pp. 1 to 35, Carnegie Institution of Washington, 1925.

Fig. 14. (*top left*) Artist Herman Beck sculpted a giant short-faced bear in the 1920s that's still present today in the park at the La Brea Tar Pits in downtown Los Angeles. Photo by author.

Fig. 15. (*bottom left*) Excavation continues at the La Brea Tar Pits, including at Pit 91, which has yielded parts of 73 saber-toothed cats, 56 dire wolves, 16 coyotes, 13 horses, 12 bison, 6 ground sloths, 6 giant jaguars, 4 giant short-faced bears, 2 camels, and a mastodon. Photo by author.

Fig. 16. (*above*) A bridge over the Klamath River in Northern California is guarded by two giant metal-cast bears on each end. Even if they're not a common sight today, bears loom large in the human psyche around the world, including in art, books, place names, sports mascots, and oral histories. Photo by author.

Fig. 17. (*above*) Paleobiologist Ron Richards of the Indiana State Museum compares the size of the giant short-faced bear skull found at Rochester, Indiana (*left*), and a Kodiak bear skull. Courtesy of Indiana State Museum.

Fig. 18. (*top right*) At the Indiana State Museum, paleobiologist Ron Richards holds the femur of a bison in his right hand and a femur of a giant short-faced bear. Courtesy of Indiana State Museum.

Fig. 19. (*bottom right*) When it was discovered in 1967, the giant short-faced bear in Rochester, Indiana, was the most complete skeleton of its kind. Casts of the bones reside in a drawer in the back room at the Indiana State Museum as well on public display there. The Field Museum of Chicago also has a display of the bear. Photo by author.

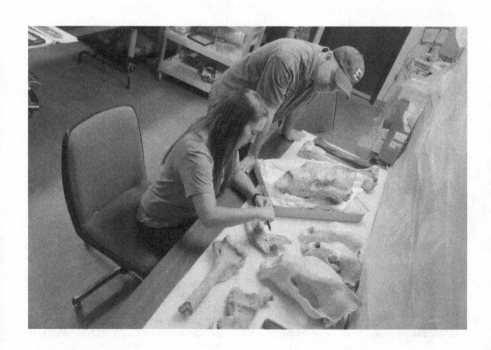

Fig. 20. (*top left*) Painter Karen Yohler's rendition of a giant short-faced bear created after the discovery of the skeleton found in Rochester, Indiana. Courtesy of the Indiana State Museum.

Fig. 21. (*bottom left*) Researchers at the Mammoth Site in Hot Springs, South Dakota, have found the remains of a giant short-faced bear that may have died after slipping into a sinkhole thousands of years ago. Photo by Isaac Zarecki. Courtesy of the Hot Springs sd, Star/Rapid City sd Journal Media Group/Lee Enterprises.

Fig. 22. (*above*) A bronze sculpture of a giant short-faced bear greets visitors at the Lubbock Lake Landmark in Texas, the site of an ancient spring popular in its time for camping and hunting. The bones of a giant short-faced bear have been found at the site, including one that appears to have been used as an impromptu tool by people. Photo by author.

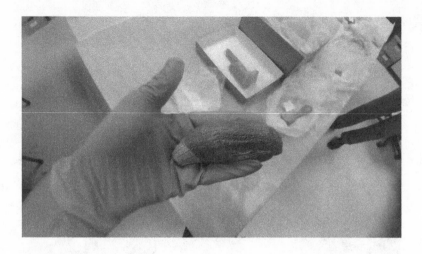

Fig. 23. (*above*) Giant short-faced bear remains found at Lubbock Lake include this canine tooth. Photo by author.

Fig. 24. (*below*) Around thirteen thousand years ago, a Columbian mammoth and a giant short-faced bear died near Utah's Huntington Reservoir. Their remains were discovered in 1988. Although they were found near each other, it's unlikely that they died at the same time. Their proximity prompted some speculation at the time that the bear may have gnawed on the mammoth's carcass. Photo by author.

Fig. 25. A comparison of skull sizes: giant short-faced bear (*top*), black bear (*middle*), and grizzly bear (*bottom*). Courtesy of the Indiana State Museum.

Fig. 26. New Hampshire–based artist Bob Shannahan created a giant short-faced bear sculpture for its *Prehistoric Giants* exhibition at the Montshire Museum of Science in Norwich, Vermont. Courtesy of the Montshire Museum of Science. Artist Bob Shannahan. Photo by Rob Strong.

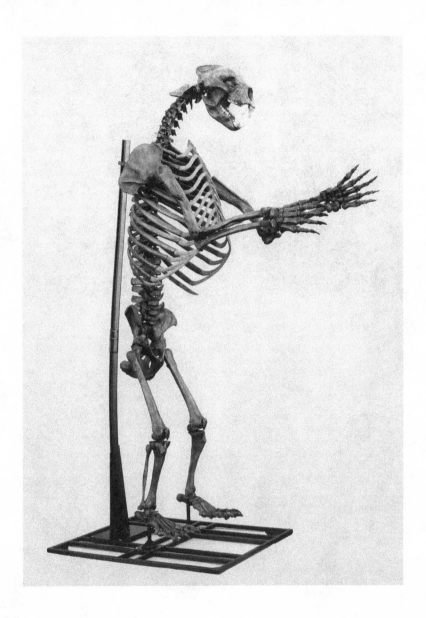

Fig. 27. Bone Clones, Inc., in Southern California produces a replica of a giant short-faced bear that lived about twenty-eight thousand years ago in what is now Alaska. Courtesy of Bone Clones, www.boneclones.com.

Fig. 28. Blue Rhino Studio in Eagan, Minnesota, created a life-size fabrication of a giant short-faced bear. Courtesy of Blue Rhino Studio.

Fig. 29. The fabrication of a giant short-faced bear produced by Blue Rhino Studio was displayed in a diorama at the Ancient Ozarks Natural History Museum in Ridgedale, Missouri, in 2017. Courtesy of Blue Rhino Studio.

PART 4

North and South

17

Fitful Arrivals

These days, if you're a paleontologist digging in a hole and think you might've stumbled upon the remains of a giant short-faced bear, the first person you're likely to call is Blaine Schubert. A native of the Ozarks, Schubert today is a professor at East Tennessee State University and director of the Center of Excellence in Paleontology. He's also an expert in the Pleistocene—and a deep expert on these long-gone bears. "*Arctodus* is one of my favorite animals," he told me. "They're amazing, terrifying, and mysterious."

There's a line you can draw about our knowledge of giant short-faced bears that starts with E. D. Cope and his initial work describing the first discoveries, then passes through the University of California's John Merriam and those looking at La Brea's earliest excavations in the early twentieth century, followed by Finland's Björn Kurtén's work in the 1960s and 1970s, and finally to Schubert in recent decades. Many others have contributed along the way—that's one of the beautiful things about science and discovery—but these scientists have provided foundational knowledge that we return to time and again to understand and challenge what we think we know about this great bear. Today

Schubert can easily rattle off details of coast-to-coast bear discoveries and offer nuanced answers about what's known and not known about short-faced bears. Like the others before him, he loves the gritty details of precision analyses and measurements alongside the imaginative side of paleontology—trying to figure out what the days were like for *Arctodus*, what they ate, how and even where they slept.

But each of those scientists had to start somewhere, and for Schubert the beginning came with an email, a grueling cave expedition in the Ozarks, and a kneecap.

In December 1998 Schubert was working at the Illinois State Museum and got an email from a caver friend, James Kaufmann, a student at the University of Missouri at Rolla. Kaufmann and other members of the school's spelunking club had been mapping a vast cave in Missouri's Pulaski County. Far inside, caver Andy Free had found the big bones of what Kaufmann suspected was an Ice Age species. It certainly could've been a cow or something else quite ordinary, Schubert thought, but what if it wasn't?

A month after the email, Schubert, Kaufmann, and a crew climbed to the entrance of the cave perched high on a bluff above the Gasconade River, a twisty strip of mountain runoff flowing through hardwood forests and along tall cliffs and brushy banks. Despite its seventy-five-foot-wide opening and high ceiling, the cave narrowed into a three-foot-tall crawlway that went on and on. Several trunks veered off to different rooms and still more cramped and twisting hallways. Cavers estimate that the underground network of byzantine passages totals more than four miles. Once inside, the trip was frigid, wet, muddy, and dark, but Schubert loves being underground. He had grown up caving in the Ozarks and elsewhere but still shivers at the memory of the trips to the bear site above the Gasconade. "Some of those trips were pretty brutal, even by caving standards," Schubert said.

Finally, more than a mile in, they arrived at a spot with a few bones on the surface and more just beneath the wet sediment. The bones on the surface were weathered and mushy from the constant moisture and exposure, but just under the surface preservation was much better. Schubert knew he was looking at a large bear, quite possibly an *Arctodus simus*, but he needed to be sure. Over the years, he'd become friends with, and was mentored by, a local and somewhat legendary biologist named Oscar "Oz" Hawksley, a retired professor at the University of Central Missouri. Bears were often the focus of their conversations, including the presence of short-faced bears in the Ozarks. In 1965 Hawksley published a paper that described the remains of at least four *Arctodus* individuals found in three separate Missouri caves, including one not far from the cave that Kaufmann had pulled Schubert into. Hawksley's collection of papers and bear bones had been donated to the Illinois State Museum. Schubert was working at the museum at the time, studying those Ozark collections.

In the wee hours of the morning, exhausted and somewhat hypothermic, Schubert exited the cave with part of the animal, a kneecap. Back at his museum, he compared it with another short-faced bear patella from Hawksley's collection and found a perfect match.

Over the next two years they returned several times to Big Bear Cave, as it became known. They worked in winter months to avoid contact with a huge population of gray bats in the cave, and the typical round-trip to the site and back lasted thirteen to sixteen hours.[1]

Once at the excavation site, the bones were uncovered, mapped, and then packed into plaster jackets to keep them safe for the journey back to the surface. It was humid, slippery, tiring work, often in extremely cramped conditions. All told, it took thirteen trips down Arctodus Avenue, as they named the passage, to get the bear out. But the result was a stunning find: ten teeth as well as vertebrae, ribs, shoulder bones, arm bones, wrist bones, leg bones, foot bones, and even hints of what may have been its hair.

Although there wasn't the right kind of material to produce a precise radiocarbon date, Schubert and Kaufmann were able to determine that the bear had been young when it died, likely four to six years old if it was a female and six to eight years old if it was a male. The bear, found on its right side, was smallish for its species and once had fractured a toe that seemed to heal just fine. The cause of death was unclear.

The discovery was also a chance to reexamine the connection between caves and short-faced bears. According to a paper eventually published by Schubert and Kaufmann, more than a third of the *Arctodus simus* remains found in the Lower 48 states were in caves, suggesting a strong connection between the two. But were the caves used by both males and females? There are two fairly helpful ways to determine the sex of a giant short-faced bear: the size (males tended to be much larger) and the baculum, which is the penis bone. So far, there's never been a confirmed recovery of an *Arctodus* baculum. One of the Potter Creek Cave specimens once seemed like a possibility, but Schubert thinks it could've come from a black bear. That leaves size as the best determinant. And because most of the other *Arctodus* remains from caves tended to be on the small side, Schubert and Kaufmann theorized that it tended to be the females who used the caves, likely for denning.

Schubert, taken by the story of the Big Bear Cave bear, offered a speculative take in the spring of 2001 about its last day outside the cave:

Imagine a day some 15,000 years ago in the region we now call the Ozarks. A deep chill is in the air as evening approaches. Winter is settling in, and the snow is beginning to pile up. A group of mastodonts is making an awful racket foraging on cold, brittle spruce around some terrace springs. On the side of a small bluff, a sabertooth intently watches a giant ground sloth and her young as they forage ever closer. A herd of flat-headed peccaries streaks across a prairie opening, having been ambushed by a pack of dire wolves. Suddenly all the animals grow tense as the birds in the forest become silent. Entering the prairie opening is a majestic giant short-faced bear. The giant grunts at the nervous wolves as she passes by, uninterested on this occasion in any food the wolves may have obtained. The bear heads up a slope and lumbers into a large cave opening. She turns to look at the snow-covered valley for a brief moment and then heads into total darkness. It will be her last look at the valley. Far back in the cave she digs out a place to bed down in the cave clay. She lies down on her right side, enters a deep sleep, and never wakes up.[2]

Since then, Schubert's been part of scores of academic studies, discoveries, consultations, and discussions about short-faced bears, but when we spoke more than two decades after that find in the Ozarks, there was still part of him taken with the animalistic feeling of simply being in close proximity to these bears.

"When I am in a cave, I travel back in time, and find skeletons, tracks and claw marks that are tens of thousands of years old right in the same clay surface that I tread," he said. "It's very easy to imagine various extinct species moving through these passages. I find peace in caves, listening and imagining the Ice Age ghosts of the past. But now imagine a short-faced bear sloshing through the cave water in total darkness, growling and breathing deeply as it approached you. That would be absolutely horrifying . . . and not so peaceful."

A few months after Schubert wrote his descriptive imagining of the fate of the bear at Big Bear Cave, another possible site was found nearby. On the morning of September 11, 2001, a crew building a new road in southwest Missouri was putting a series of explosive charges in the ground to blast away a large rock outcropping. Then a confusing call came in. There'd been a terrorist attack in New York City, chaos was unfolding, and the full scope was unclear, so there shouldn't be any blasting. The loaded explosives, though, couldn't be removed from the holes, so the crew got special permission to detonate them but do nothing further.

The situation only got stranger. The result was no mere crater in the limestone. It peeled part of the roof off a massive underground cavern. There are thousands such caves in Missouri, so the crew knew the drill. The road work was called off and the karst experts were called in.

That's when Matt Forir's phone rang. He was a graduate student at what was then Southwest Missouri State University working under Ken Thomson, one of the state's most prominent experts on caverns, especially in the Ozarks. "I was glued to the TV," Forir later told *Rural Missouri*. "The phone rings and it was Ken Thomson saying this new cave opened up. I wasn't interested. I was focused on the TV."[3]

Still shaken by the day's events, Forir and his caving partner Lisa McCann did their duty. He expected to find just another trash-filled hole, but his expectations quickly went out the window. "The first thing we found was a claw mark on the wall," he said. "It was not just a claw mark. This thing is eight inches wide, two feet long. It was twelve to fourteen feet off the ground."

It would soon be clear that the cave they'd stumbled onto was more than just a cave. It was a portal into another world and another time, one that held forgotten primeval beasts at once foreign and familiar. Researchers would later estimate that the long-sealed two-thousand-foot cavern network south of Springfield held sediment more than one million years old. Amid the drip-shaped stalactites and soaring cathedral ceilings, the cave was an open book of thousands of years of life, death, decay, preservation, and drama. Claw marks, bones, tracks, and dung. "This site," one breathless investigator later said, "is really one of the most important Ice Age scientific sites on planet Earth."[4]

After its discovery, the cave was sealed for several months so the road could be rerouted and built, and to develop a plan for exploration. Word leaked out, though, and vandals found their way in. A calcite column was carved, skeletal remains of a snake were smashed, and some of the mineral formations were broken off and stolen. It was ultimately protected, and researchers spent years investigating Riverbluff Cave, as it had been named. They found evidence of wolves, peccaries, mammoths, and tortoises. Some may have lived in the cave; others may have been dragged inside.

And the claw mark seen so high up on the wall that very first day? Forir speculated it came from a giant short-faced bear, perhaps ten feet tall on the hind legs, reaching over its head to scratch deep grooves into the red and white walls of the cavern. "That split second of that animal's life is preserved in this cave," Forir told *Rural Missouri*. "And we can identify, we can see what it was doing. And you never get that from the bone." One of the most obvious benefits of the scratch marks is a simple demonstration of just how big they could be. Not to mention imagining what those blade-like claws might do to any flesh it could find. "The short-faced bear is the T. rex of the Ice Age," Forir said.

Like the spelunkers' find at Big Bear Cave or the cave inadvertently exposed by the roadside blasting crew, I've often been struck by the chance nature of these discoveries, especially paleontological remains. A worker happens to spot something amiss or scrapes away the top levels of sediment, often for utterly utilitarian purposes, and exposes the remains of some extraordinary being hidden away for thousands of years. Or someone stumbles on a pile of bones on a hike or a stroll on the beach or while clearing the way for a new road. Over the past century, the giant short-faced bear has arrived in the same fitful and irregular fashion, especially in recent decades. And each added another brush stroke or two to the emerging portrait of *Arctodus simus*.

In 1982 Sam Baker, Dennis Hone, and Robert Swain, all members of the High Desert Grotto club in eastern Nevada, were exploring a cave more than 6,700 feet above sea level in the Schell Creek mountain range, not far from the dry, scrappy mining town of Ely. Labor of Love Cave, as it was soon named, extended about 450 feet into the mountain with a small stream that flowed the entire length. Near the end of the cave, in two pools about thirty feet apart, they also discovered ancient bear bones. A huge scattering of them, some under water on the pool bottom, some wedged into the surrounding wet silt. All told, there were remains of at least three black bears and two giant short-faced bears, all from the late Pleistocene.

Worried that the remains might be stolen by thieves or washed away, the U.S. Forest Service soon issued an emergency permit to collect and preserve them as soon as possible. That summer Steven D. Emslie, then with the University of Florida, began photographing and mapping the bones and then wrapping them in aluminum foil for transport. It was a complicated scene. At the first pool, aptly labeled Bone Concentration No. 1, all of the bones were submerged or partially buried in silt at the bottom. That collection included an immature giant short-faced bear and a fairly complete skeleton of a black bear. At Bone Concentration No. 2, the bones were both on the pool bottom and on the pool's edge. That area included a nearly complete skeleton of a small adult giant short-faced bear. In both places the bones lay in clusters, skull fragments next to leg bones and mandibles and broken arm shafts and vertebrae. Teeth were here and there, as were bits of chipped bone. Each item was painstakingly wrapped in foil, placed in a box, taken out of the cave, and driven to the Natural History Museum of Los Angeles County, the same entity that runs the La Brea Tar Pits.

In 1985 Emslie and Nicholas Czaplewski, then at Northern Arizona University, published a paper about the discovery, trying to piece together the lives of these ancient bears. They speculated that the late Pleistocene habitat that these bears roamed in eastern Nevada was likely open, sagebrush-covered flats with a scattering of lakes in the valley bottoms, bristlecone forests on the valley edges, and surrounding mountains with alpine glaciers and high-elevation vegetation. The situation inside the cave seemed to indicate the bears may have died there and some of their bones were either scavenged or washed away.

"The relative completeness of the black bear skeleton from BC-1 and the short-faced bear skeleton in BC-2 suggests that these animals died in the cave, perhaps during a period of winter hibernation," the scientists wrote. "Bones of other animals found in the cave also display breakage and tooth marks, and may represent food items transported to the cave by carnivores."[5]

Emslie and Czaplewski also took the opportunity to closely examine the nearly complete *Arctodus simus* skeleton and poke at Björn Kurtén's hypothesis that the giant short-faced bear was a highly predacious, flesh-eating monster. Instead, they offered, the bear was "probably largely herbivorous, but retained bone-crushing capabilities and may have been an opportunistic predator and scavenger." That story, they said, was told in the bear's skull, teeth, and limbs.

With the skull, Kurtén years before had maintained that over-sized features like the arching dome, huge nasal cavity, and orientation of the cheek bones converged to form the perfect machinery for a carnivore. Emslie and Czaplewski countered that those may have simply been the result of the bear's large size—and noted that the same features were obvious in the South American spectacled bear, the giant short-faced bear's closest living relative, which has a well-established reputation as primarily a vegetarian with a yen for munching mostly on bromeliads, leaves, bark, and fruits (with occasional bits of carrion). They also said the teeth of A. *simus* suggested a largely veggie diet. While Kurtén said the imposing canines and high-crowned nature of the first molar made it obvious that it was a predator, Emslie and Czaplewski saw it differently. Although the teeth and jaw were well built for occasional bone-crunching, they found it notable that the molar cusps of giant short-faced bears tended to get flat and worn with age. "The general trend in most highly predacious carnivores is to maintain a sharp, shearing blade on the carnassials," they wrote. They also challenged the idea that the giant short-faced bear's long, slender limbs had evolved to chase down and kill other animals. If the bear was more of a vegetarian, they said, those long limbs may have aided in traveling through high bushes and grasses in open lands, or possibly in tearing down shrubs and small trees to feast on leaves, fruits, or bark.

There were other finds, other pieces of the puzzle to place.

In Utah in 1978, south of Salt Lake City, a heavy equipment operator named John Yackovich was loading sand and gravel from an industrial pit when he found the bones of a giant short-faced bear. A 1983 study of the remains estimated it had been a hefty male, likely weighing around 1,450 pounds and possibly more in a fattened state before the winter. The bear lived around 14,500 calendar years ago, and his bones ended up buried in the sand and gravel along the far southern shore of Lake Bonneville, the sprawling prehistoric lake that once covered nearly 20,000 miles and was a predecessor of the Great Salt Lake and its southern sister, Utah Lake.

The remains—which included the right thigh bone, two damaged shin bones, three vertebrae, and pelvis fragments—were the first evidence of *Arctodus* in Utah. It was, the researchers said, "a very large mature male," likely closer in sheer size to the gigantic one found by the gold diggers in the Yukon in the early 1900s and certainly outsizing those from La Brea and Potter Creek Cave.[6]

In the early 1990s, as a new housing subdivision was being built in Murrieta, California, in the southern part of the state, a paleontological crew was called out to look for ancient bones. They weren't disappointed. During 1990 and 1991, inspectors found more than thirty thousand vertebrate fossils from the Pleistocene at thirty-six different spots. They included the remains of extinct horses, mammoths, antelope, and camel. Also in the copious collection of Riverside County bones were the left heel and ankle bones of a large, male giant short-faced bear.

A more dramatic story unfolded farther north. In January 2010 a man operating an earth scraper in a potato field near eastern Oregon's La Grande Airport looked back to see a patch of white in the silty, gray sand in his wake. He got off the tractor to investigate and found the broken shinbone of a mammoth in the dirt. Over the next two weeks of digging, more bones were found in the same area (about sixteen feet below the surface), including mammoth tusks, parts of a prehistoric bison, and the broken knobby end of the thigh bone of a giant short-faced bear. Pleistocene fossils had been found before near La Grande, like a ground sloth from about 12,900 calendar years ago discovered along a roadside and the remains of a Columbian mammoth, dated to around 18,000 calendar years ago, at Eastern Oregon University. The bear was something new, though. Researchers determined that it was likely an adult *Arctodus* about the size of those found at Potter Creek Cave and La Brea. But the find posed something of a mystery. All three animals had been entombed in a sandy bed fairly close together about 15,000 calendar years ago. What happened? Scientists used pollen, freshwater diatoms, and other minuscule discoveries at the site to get a better picture. Grasses and pondweed growing on the banks of nearby lakes and creeks might've been food for mammoths, bison, and other herbivores, they determined. Junipers likely dotted the landscape and hemlock covered the surrounding hills. The bodies were found in an area that was once a shallow stream and was later inundated with a huge pulse of sediment that fanned out. The end may have come in a spasm of natural violence. "They may have drowned in a spring flood . . . released when an ice dam high in the mountains burst or they may have been killed and butchered by late Pleistocene hunters," researchers wrote in a 2014 article for *Oregon Geology*.[7]

In 2008 a man named Phil Graydon found a bone on a beach at the base of a steep cliff a few miles north of Victoria, British Columbia. It turned out to be the fragment of a forearm bone of a large adult giant short-faced bear. Radiocarbon analysis revealed the bone to be about twenty-seven thousand calendar years old, researchers said later. It was "markedly more robust" than the largest brown bears in the collections at the Canadian Museum of Nature and close in size to the big extinct bear found in the Yukon a century earlier. But it raised an interesting question: How did it get there? The bone was found on the far southeastern point of Vancouver Island, the twelve-thousand-square mile island that today sits among a scattering of smaller, densely forested islands on the west coast of Canada. (Other extinct animals found there include mammoth, mastodon, and helmeted musk ox.) About twelve miles away, on San Juan Island, the jaw of a giant short-faced bear had also been discovered. In 2010 Martina L. Steffen of the Royal BC Museum and C. R. Harington of the Canadian Museum of Nature offered an intriguing possibility: "Early occupation of the islands by bears could have been by way of ice or land connections that may have been present for brief periods of time as ice advanced or receded through this region, or they could have traveled by swimming, perhaps by way of the Gulf Islands or San Juan Islands as stepping stones," they wrote.[8]

The idea of ocean-going swimming bears isn't so outlandish. In the Arctic, polar bears regularly swim thirty miles or more—and one female swam some 426 miles over nine days, according to a 2011 study. Giant short-faced bears wouldn't have been alone in the Pleistocene in terms of large beasts paddling through the sea. Some believe that Columbian mammoths, following the scent of food, swam to the Channel Islands off the Southern California coast, where they subsequently evolved into smaller versions of themselves, the delightfully named pygmy mammoths.[9]

And then there were the *Arctodus* bears found in Florida. For years, Kurtén and other scientists had noted the lack of giant short-faced bears in the southeastern United States, speculating that the region may have been ceded to its smaller cousins, *Arctodus pristinus* and *Tremarctos floridanus*. In 2010, though, Blaine Schubert and other scientists announced they'd found the remains of at least three giant short-faced bears in central Florida, all within about six miles of each other. One set was found along the Rainbow River and more were found at Lake Rousseau. The finds increased the known range of giant short-faced bears by hundreds of miles to the southeast and offered another piece of evidence that these bears were highly adaptable, not only capable of living in forests or open landscapes to the north but also in the subtropical woodlands and savannas of the region.

Although it's not clear from the fossil record whether *A. simus* and the smaller *A. pristinus* ever coexisted, it's logical to assume they each competed for food alongside a suite of other meat-eaters there, like dire wolves and saber-toothed cats. No doubt there were some dramatic showdowns. "The much larger body size of *A. simus*," the scientists wrote, "would have provided an advantage in disputes over carcasses."[10]

Indeed, these were massive, well-built animals hardwired, like all of us, to relentlessly pursue survival in a difficult world. We're still puzzling over exactly how they did it.

18

Imagining *Arctodus*

Close your eyes and imagine a bear.

You probably see a burly mass of fur, fat, and muscle—maybe shaped like a giant potato—with huge teeth, massive claws, and friendly, round ears. Your mind might also latch onto what this bear does. It's a powerful, hungry meat-eater that might have an ornery streak and is apt to exterminate any smaller animal within sight and make a quick meal out of it. It's speedy despite its size. Cunning. And always on the lookout to kill some fresh meat.

That's not altogether wrong. But not really right either. It depends. Polar bears, for instance, are excellent hunters that dine mostly on ringed and bearded seals. Panda bears eat bamboo, almost exclusively. Grizzly bears' diets vary greatly, often based on where they live and what's available. At Yellowstone National Park in Wyoming, they like elk carcasses, trout, nuts, and migrating moths. A few hundred miles north at Glacier National Park in Montana, they tend to favor more meat and berries. How a bear acts and what a bear eats is determined by a lot of factors, including its environment, the weather, competition, its size, shape, and even the architecture of its bones. Even with the advantage of observation in the field, mysteries abound about the lives of bears in our midst.

Trickier still is reconstructing the life and behavior of a bear you've never seen, divining the course of its days and habits from bones, teeth, collagen, the contextual materials at the discovery site, and what we know about how extant animals behave. Conjuring a dead thing into imagined existence is the business of paleontologists and other scientists, a sort of informed speculation that provides fuel for debate at conferences, bars, and academic papers alike.

When Finnish paleontologist Björn Kurtén was describing giant short-faced bears in 1967, he felt certain that, after his examination of several dozen specimens, a fairly vivid portrait had emerged. These were large, fast, carnivorous beasts capable of bringing down some of the Pleistocene's megaherbivores, like mastodons and mammoths. He based that idea largely on the bear's long legs and front-facing feet, which allowed it to reach top speeds, not to mention its huge cranium and jaws, which seemed perfectly designed for delivering a fatal bite to the land's mightiest residents.[1]

But he wasn't the first, of course.

E. D. Cope, in his 1891 paper, reexamined the Potter Creek Cave skull and deemed it "the most powerful carnivorous animal which has lived on our continent,"[2] and a decade later the *San Francisco Call* was declaring it the "strongest, fiercest animal of America."[3] In subsequent years, Merriam, Furlong, Sinclair, and Lambe in Canada did their part to build the monster from the bones they'd found. Kurtén's 1967 assessment helped cement the reputation of *Arctodus simus* as a sort of super-predator, a notion with such power that it's carried into the present day.

But science is meant to be challenged, especially when it comes to reconstructing lives long since gone.

Recall that the scientists—Emslie and Czaplewski—who examined the bear remains from Nevada's Labor of Love Cave in 1985 were pretty sure Kurtén had gotten it wrong. *Arctodus*, they said, was "probably largely herbivorous," as evidenced by worn teeth that indicated heavy chewing of plants. It may have dabbled in meateating either as a predator or scavenger, but that likely wasn't its main diet. Rather than being built to be a mighty hunter, *Arcto-*

dus's tall and slender build may have actually been an adaptation for grabbing foliage or berries off high shrubs or to have a higher vantage point above tall grass, they said.[4]

The truth may have been somewhere in the middle, between Kurtén's super-predator and the vegetarian-leaning bear proposed in 1985. In a series of publications in the 1990s, Paul E. Matheus, a University of Alaska Fairbanks paleontologist, attempted to get to the bottom of it by taking a deep dive into the locomotion and ecomorphology of giant short-faced bears.

Matheus first looked at stable isotopes found in the collagen of giant short-face bears who lived in Alaska around twenty thousand years ago. These bears, he determined, loved meat and ate a lot of it, but, unlike brown bears in the same region, they didn't seem to eat salmon. Still, knowing these bears had a meat-heavy diet, the question turned to: How did they get it? Were these huge bears really capable of chasing down prey?

In puzzling over the "*Arctodus* problem" of locomotion, Matheus cautioned against trying to compare short-faced bears with other bears or even other members of the carnivora family, extinct or otherwise. These bears—so large a body atop long, slender legs—weren't simply super-sized versions of black, brown, or polar bears, nor were they really catlike in their entirety. It's as if they'd been the designed with the parts of several different kinds of species, one of nature's many mash-ups. "There are no good analogs for *Arctodus*," Matheus wrote.[5]

Being a great predator isn't just about running fast. It's also about rapid acceleration and the ability to stop on a dime during a chase, turn another direction, and stay in the hunt. Not all prey run in a straight line, and predators need to be able to zig and zag along with them. To determine whether giant short-faced bears were capable of that kind of speed and maneuvering, Matheus exhaustively compared the strength and thickness of *Arctodus*'s front and hind leg bones to its closest relative (the spectacled bears in South America) as well as other bears in North America. He found that *Arctodus* bones tended to be lighter and thinner (relative to size) than many other bear species, especially their front

limbs, which were longer than their hind limbs. That fact, compared with its large size, meant that "it does not seem plausible
that this bear was built to withstand the high forces of fast running and predation." Simply put, the limbs of a giant short-faced
bear were likely to snap under its heavy load if it attempted to suddenly stop or take sharp corners during a high-speed chase. Those
kinds of injuries could prove fatal.

In other ways, too, giant short-faced bears didn't move like most
other bears, Matheus said. They didn't plod and amble in the awkward way that, say, pigeon-toed grizzlies seem to do. They paced
with efficient swinging legs. With front-facing feet, shorter hind
legs, short backs and longer front legs, they may have moved more
like some hyenas. Matheus contended they were built for long-
distance, medium-speed travel, perhaps around 8.5 miles per hour,
with short bursts of speeds capable of hitting around 28 miles per
hour. (That's not quite as fast as, say, grizzly bears, which have been
known to hit up to 35 miles per hour.) *Arctodus*, Matheus wrote,
"did not evolve as a powerful super-predator, but rather as a lanky,
far-roaming bear which I propose was a unique scavenging spe-
cialist on Pleistocene landscapes."[6]

That's not to say that giant short-faced bears didn't sometimes
chase, catch, and kill prey, Matheus said. Undoubtedly they did
so when there was an easy target, he said. But that may not have
been the primary way they lived and evolved. Instead, he theorized that *Arctodus* moved over vast ranges with its fast-traveling
gait, following its nose to dead mammoths and other carcasses. Its
expertise may have been kleptoparasitism, the fine art of feasting
on animals that others have killed. And, he said, it probably used
its finite bursts of speed to scare other animals off carrion or flee
if suddenly outnumbered by other hungry predators.

"I suggest that this carnivorous bear had evolved as a specialized scavenger adapted to cover an extremely large home range
in order to seek out broadly distributed large-mammal carcasses
and to dominate this lucrative, but unpredictable carrion source,"
Matheus said.

Around that time bear biologist David Mattson offered his own reexamination of the bodily traits of the giant short-faced bear, primarily by applying what he knew about extant bears to speculate about the lives of *Arctodus simus*. By and large, he didn't buy the ideas that came out of the Labor of Love Cave discovery. It didn't make sense, he said, that its long legs were meant to provide a better view over the ground cover, especially when most of the grasses and other cover during the late Pleistocene tended to be less than a meter tall across what is now the western United States. Also, Mattson said, the bear's short snout and flat face would've been a decided disadvantage in trying to procure flowers, seeds, leaves, and berries from trees and shrubs. Its skeleton seems to indicate a relatively mobile bear with a penchant for carnivory and a skull capable of killing and dismembering prey: "If this bear engaged in ambush-type predation, it would likely have been swift enough to catch its large-bodied prey, perhaps grasping it from the rear with its paws and collapsing the hind-quarters with a combination of weight and a crippling bite to the back, much like contemporary brown bears." He didn't discount Matheus's hypothesis that short-faced bears were mostly scavengers rather than predators but noted that spotted hyenas and brown bears today are often referred to as scavengers, even though they can be "formidable predators and can, in fact, derive much of their meat from kills."[7]

In 1996 three other scientists offered their own, slightly different version of *Arctodus simus* after examining the skull, jaw, teeth, and other parts of a young adult that died in a sinkhole at what is today Hot Springs, South Dakota. The sinkhole formed when the ground collapsed and filled with water from an underground spring. Animals wandered into the pond and then found themselves stuck when they couldn't climb up the steep sides. So far, more than sixty mammoths have been excavated there—it's now called the Mammoth Site—along with other animals, including a short-faced bear that may have been attracted by the smell of death and then drowned or suffered a blow "from a not-so-dead mammoth."[8]

Looking at the size, dimensions, and morphology of the bear from Hot Springs, the three scientists—Gennady Baryshnikov, Larry D. Agenbroad, and Jim I. Mead—theorized that *Arctodus* was a far-ranging scavenger adapted for the plains and grasslands. It was highly opportunistic and drawn to death traps for prey like the La Brea Tar Pits, Natural Trap Cave in Wyoming, and the sinkhole in South Dakota. "This was an omnivorous bear, which consumed plant foods and preyed on juvenile and sick ungulates, but primarily scavenged the carcasses of dead mammoths and other large mammals," they said, later adding: "*Arctodus simus* probably ranged with the herds of mammoth, bison, and other large herbivores looking for the converging of carrion-feeding birds (vultures, condors, teratorns) as a clue to carcass locations."

They speculated that *Arctodus*, unlike grizzlies or European cave bears, may not have hibernated. Winter is often deadly for herbivores, and short-faced bears may have stayed awake to take full advantage of all of the carcasses in the coldest months.

Boris Sorkin at Northern Illinois University offered his own take in a 2006 paper in the journal *Historical Biography*. He tended to agree that *Arctodus* was mostly a traveling scavenger and argued some additional points to discount the idea that it was an "active predator." For one, he said, the upper canines of giant short-faced bears were short relative to the front-to-back measurements of their skulls, which would have made it difficult to deliver a fatal bite to the trachea or spinal cords of large prey. Also, he said, the eye openings were small compared to, say, lions or other Pleistocene cats, which likely meant they had poorer eyesight than their felid competitors. Further, the bears' arm bones and muscle makeup would have made it difficult for them to sneak up on prey and quickly accelerate from a crouched position, another shortcoming when compared to big cats. That said, Sorkin had his doubts "whether a nonflying animal could subsist on scavenging alone, even in an environment such as the Pleistocene North America in which large animal carcasses were available year-round, because of the high energy cost of searching for animal carcasses on foot." For him, that raised the prospect that plants remained a key part of the giant short-faced bears' diet.[9]

In 2010 a group of scientists from the University of Malaga in Spain looked to further reconceptualize the bear with a paper in the *Journal of Vertebrate Paleontology* called "Demythologizing *Arctodus simus*: The 'Short-Faced' Long-Legged and Predaceous Bear That Never Was." It wasn't that they questioned the existence of this extinct bear, but they wanted to dismantle some of the lore that had calcified around it over more than a century. Through a series of measurements and comparisons, the authors contended that the giant short-faced bear was perhaps more like other bears (though larger) than had been portrayed. The snout of *Arctodus simus* may not have been as short as some speculated, and its legs were probably not as relatively long as earlier presumed, they said. The bear's daily behavior and eating habits varied depending on where it lived and in places like the Far North was probably similar to wide-ranging brown bears living in the region today, which both hunt and scavenge, dining often on flesh and bone marrow. The bottom line, they said, was that it may be time to do away with the concept of the bear as a short-faced, long-legged super bear.

"Our results do not support the previous views of *A. simus* as a fast-running super-predator or as a specialized scavenger. In contrast, the picture that emerges from this study is one of a colossal omnivorous bear whose diet probably varied according to resource availability," they said, adding later: "The current situation demands a deep taxonomic revision."[10]

Still another group of scientists in 2013 tested the hypothesis that giant short-faced bears were "hyper-scavengers" that regularly crushed bones in their teeth during a meal. They focused solely on reconstructing the bears' diets through an examination of the texture of "dental microwear," which allowed them to determine "evidence of actual food choice during the last days or weeks of life." The study, limited to just *Arctodus simus* remains at La Brea, found none of the telltale signs on their lower molars that their teeth had been used to regularly consume bone. Again, while meat was almost certainly a part of the diet around La Brea for these bears, the wear on the teeth indicated that plants were in the mix too.[11]

After poring over several decades of ideas and interpretations, I couldn't help feeling as if the remains of these bears made something of a Rorschach test, inky shapes delivering a slightly different meaning to each beholder and the truth remaining just out of reach. So where does that leave us when trying to paint a picture of the life of a giant short-faced bear in the Pleistocene?

I put the question to Mattson, the bear biologist who wrote the paper in the 1990s parsing what Matheus had said as well as the speculation that spilled from the finds at Labor of Love Cave in Nevada. I'd crossed paths with Mattson several times during my work covering grizzlies in Yellowstone and the Northern Rockies. He spent decades studying bears and was an ardent voice for protecting grizzlies under the Endangered Species Act. We'd been on a few hikes together, including one memorable one on a ridge high above the Yellowstone River when he pointed out a yellowish, sandy patch near the water where he said bears would come in the spring to eat the soil at that particular spot because it was rich in potassium, something that was often lost during hibernation. I remember feeling like I'd been let in on a secret about a lost corner of Yellowstone, something that maybe only Mattson and the bears knew about.

"It's the nature of scientific publication that everyone claims to have the last word," Mattson said when I called to ask about what to make of so many views about the lives of *Arctodus*. "And that holds for the short-faced bear."

These were not vegetarians, he said, but their eating habits were highly influenced by where they lived and what was happening at the time. "It's an essential trait of all bears that they are versatile when it comes to diet. You have to believe that the short-faced bears were capable of some dietary versatility."

And of course it makes sense that, because most bears are such expert opportunists, *Arctodus* likely ate what was available and made do in its environment. In the Far North, that may have meant a steady diet of caribou. Farther south, perhaps it was peccaries or dead mammoths. In today's Southern California, the menu may have relied a bit more on plants. Whatever the case, these giant bears needed, and found, lots of calories to keep their bodies running and their brains firing in a dynamic world full of dangers and changes.

Clearly these bears were wanderers. Mattson speculated that giant short-faced bears may have had home ranges that topped a thousand square miles, depending on food availability and the presence of competitors. Occasionally they certainly ran into people too, Mattson said, which must've been a shock. "*Arctodus simus* stood out as being truly enormous. I think just the sheer size and the fact that it's a bear, people immediately go to 'holy shit,'" Mattson said with a laugh, adding that eventually humans' fear turned to fascination. "That's the part that goes right to the brain stem: they're just so huge. It piques people's curiosity. How did a bear like that make its living?"

Blaine Schubert summed it up to me like this: These were large, omnivorous bears that could travel great distances and eat a wide variety of food. In some places, like in the Far North, they likely ate a lot of meat. Farther south, like around La Brea, they seem to have had more plants in their diet. But you can't necessarily apply data from one region—like how much meat was in its diet—and assume that applies to bears living somewhere else. They're too adaptable, too wily, too omnivorous. They could hunt, they could scavenge, they could scare other animals off a carcass, they could dine on plants. Their bodies, especially those long legs, allowed them to travel great distances—answering the whims of desire or the scent of food on the wind. One thing is clear, though. The bears, especially the tallest and mightiest, were often able to call the shots.

"They could eat whatever they wanted," Schubert said.

19

The Great and Far North

On the afternoon of August 3, 1979, geologist Alan Morgan stepped off a helicopter in a bend along the Ikpikpuk River in a vast, treeless delta about ninety miles north of Alaska's Brooks Range. Morgan was a geologist with the University of Waterloo in Ontario, Canada, but that summer he had a temporary job on Alaska's North Slope working for the U.S. Geological Survey. He and his crew were mapping part of Naval Petroleum Reserve No. 4, the area today known as the National Petroleum Reserve, which sprawls across some twenty-three million acres of Alaska's North Slope.

While the other two researchers picked their way along the riverbank, Morgan was at the water's edge looking for beetles for his own research. Soon he found broken fragments of a mammoth tusk, and then something more intriguing as the weather worsened. "At the point that we were ready to leave I noticed a large bone just ahead of me. Since it looked identifiable I picked it up," Morgan, now retired from the University of Waterloo, told me. "The bone was hefty, and about a foot and a half long. I can remember thinking this was very much like a bone that an earlier human might have used as a club. The leg bone was chunky, very heavy, and would have made an ideal weapon in an area where trees were absent."

It was mottled light and dark brown with a few streaks through it—and well preserved, likely from being entombed in the permafrost. Morgan had no idea the bone was from a bear, although he knew there were big brown bears in the area. "We were well prepared for encounters," he said. "Fortunately, we only saw them in the distance from the air."

The leg bone came back with him to Ontario and sat in his office for years at the University of Waterloo. One day a visiting paleontologist working with the Royal Ontario Museum walked past and poked her head into his office. "Wow, that's a really interesting bone!" Morgan remembered her saying. "Would you mind if I take it back to the ROM?"

Scientists ultimately determined it was the left upper arm bone of a giant short-faced bear. It probably belonged to a young male. "The bone is massive, despite its immaturity," scientists wrote in a 1993 paper about it. From the bone's length, which was about 18 inches without the rounded end that fits into the joint, researchers estimated that the bear may have weighed more than 1,100 pounds. Radiocarbon dating estimated the bear lived around 27,000 years ago.[1]

Morgan's find also provided a chance to freshly speculate about exactly what this bear looked like and how it moved. Although the humerus was long, it was also more slender than those of other large bears. The bone's dimensions and articulation were "suggestive of a straight, stiff-legged, swinging gait rather than the gathered, flexing and straightening motion of other running carnivores and ungulates," the paper said. The giant short-faced bear didn't run like other bears, and that meant it may not have acted like other bears either.

But there was a greater importance to the discovery along Ikpikpuk River. It significantly expanded the known range of giant short-faced bears to the upper reaches of the continent. Remains had been found around Fairbanks, Alaska, and the Yukon but never so far north. And it planted *Arctodus simus* smack in the middle of the continent's most important corridor in terms of the arrival of wildlife and people (at least some of them) to North America: Beringia.

One summer in my early thirties I flew into the desolate north coast of Alaska, landing on a narrow airstrip not far from a crook in the indolent Sag River. The town of Deadhorse, at the farthest northern edge of the continent, is deep in the Arctic Circle and existed essentially as a few clusters of industrial, no-frills buildings that sprang up in the Prudhoe Bay oil rush in the late 1960s and 1970s. I arrived in June to visit a climate research station about 130 miles to the south on the edge of the Brooks Range. It was part of a journalism fellowship with the Marine Biological Laboratory, and I was keen to plunge into one of the most remote places I'd ever been.

It didn't disappoint. Once we piled into a van and left Deadhorse and its trucks, stacks of pipes, and piles of industrial equipment behind, the North Slope revealed its real and surreal self, a mix of treeless tundra, spongy moss and lichen underfoot, layers of permafrost farther down, steep and often barren mountains, icy streams, and blue-hued snow fields. The never-ending summer daylight and frenetic hordes of mosquitoes added a surreal element. There were also musk oxen, caribou, Dall sheep, snowshoe hares, Arctic char in the frigid streams, and more than a few irresistibly cute, pointy-eared red foxes.

I spent the next week or so exploring this otherworld and talking with scientists based at the Toolik Field Station about their work tracking and understanding climate change at the northern end of the planet. Bears, though, were never far from my mind. I badly wanted to see one and was constantly sweeping the tundra with my binoculars in search of a silhouette with stubby legs and a big, rounded rump. I'll admit my enthusiasm for a bear encounter lessened during a mountain hike as we picked our way through bushy thickets taking turns to shout, "Hey, bear! Coming through, bear!" I wanted to see a bear, not run into one in a surprise encounter.

Alaska is the Land of Bears if there ever was one. A few thousand polar bears occupy the northern edge of the state. Elsewhere there are about thirty thousand brown bears (also known as grizzlies, especially in the Lower 48 states) and some hundred thousand black bears. There are stuffed bears in the airports, bear images on signs everywhere, and the business of viewing the famed brown bears of south central Alaska, the largest concentration in the world, generates around $34 million in sales.[2]

It seems no accident that the word "Arctic" has its Latin (*arcticus*) and Greek (*arktikos*) roots in bears, referring to Ursa Major, the bear-like constellation in the northern sky that includes the Big Dipper. And it's no coincidence that the Big Dipper is featured on Alaska's state flag.

One afternoon we trekked across the soft and squishy tundra, doing our best to keep the mosquitoes from settling on us for long, to the edge of a stream. Most of it was lined with gravel, but one spot had been exposed and in the brown mud there was a single paw print: the four-inch-wide pad from the front of the foot and, just above it, five toe indentations, all neatly in a line. My body tensed and I looked up thinking that the brown bear was still around but quickly realized it was probably long gone. That's as close as I got to a bear on that trip (as far as I know), but it was hard to shake the feeling of some ancient and fleeting connection, a single human crossing paths with a single bear in a primeval place where people and beasts have breathed the same air, and traveled the same Far North ground, for thousands of years.

Toward the end of the Pleistocene, around 70,000 years ago, much of the world was experiencing a profound cold snap. Large swaths of North America were groaning under the weight of glaciers, some nearly two miles high, and stretching across millions of square miles. The situation was much the same in other parts of the world. In fact, so much of the planet's water was locked up in ice sheets that sea levels dropped as much as 300 feet in some places. As the oceans retreated, a huge belt of the seabed was exposed between Siberia and Alaska. The wide, shallow platform known as the Bering land bridge stretched some 1,000 miles from north to south, the connecting corridor between Asia and North America, Old World and New. The land bridge, which lasted until about 25,000 years ago, was only part of what's called Beringia, a vast and complicated area that once stretched from eastern Siberia, through much of Alaska into the Yukon Territory.[3]

This wasn't quite the Alaska we think of today, with its endless miles of squishy peat, permafrost, tundra vegetation, vast boreal forests, and layers of snow on the ground for much of the year. You could be forgiven for assuming the Pleistocene's frigid conditions had turned Beringia into a singular, bone-chilling snow drift where life for many animals would've been next to impossible. Not so, though. Beringia was typically a raw, ice-free landscape where temperatures were too cold for trees to grow. Clear sunny skies and dry conditions promoted steppe-like conditions where grasslands flourished and the ground remained firm enough for hooved animals to walk around throughout the year without getting stuck in muck or deep snow.

As the oceans receded, grasses, mosses, and lichen grew on the exposed land bridge, enticing woolly mammoths, mastodons, bison, caribou, horses, musk ox, and mountain sheep to wander from eastern Siberia and into the heart of Alaska, the Yukon, and beyond. And wherever herbivores roamed, predators followed, including brown bears, gray wolves, American lions, and scimitar cats (apex predators related to saber-toothed cats). Giant short-faced bears were there in plentiful numbers too, though it's likely they evolved much farther south and had wandered north into Beringia only to have giant ice sheets close in behind, isolating them from the rest of the continent for thousands of years.

Whatever the case, Beringia blossomed into a lively place with shifting climate and weather conditions, ebbing and flowing vegetation, vast movements of herbivores, and dynamic relationships between predators, prey, and the landscape. Adding to the drama was the arrival of weapon-wielding humans—families and larger groups who arrived from the west and found a foothold on the cold landscape, hunting what they could, traveling when they needed, and doing their best to survive.

Paleontologists are still trying to piece together the picture of animal life in Beringia. In 2008 scientists from the Smithsonian Institution, University of California—Santa Cruz, and Uppsala University in Sweden published a paper attempting to reconstruct the diets of late Pleistocene carnivores and herbivores whose remains were found near Fairbanks, Alaska. Looking at stable carbon and nitrogen isotopes from bone collagen, they determined that horses, bison, and mammoths primarily ate grasses, sedges, and herbaceous plants, while caribou and musk oxen went for lichen, fungi, and mosses. Meanwhile, the predators tended not to be very picky. Most sampled all the herbivore species that were available, with wolves showing the widest selection of meats. The few giant short-faced bear samples that were analyzed indicated a particular taste for caribou and a limited dabbling in mammoth—a little surprising since they were among the only predators thought big enough to bring down a mammoth on their own (the lions and cats might've been capable too).

The scientists also noted it's unlikely that all five of those top predators were on the stage at the same time and place. Although wolves seemed to persist throughout, the scimitar cats and the American lions didn't appear to be operate simultaneously—and neither was around during the time giant short-faced bears were present, roughly between 40,000 and 20,500 radiocarbon years ago. "Thus, short-faced bears either dictated which carnivore species were present or were able to persist locally while other species were absent," they said. In fact, they noted that the sudden disappearance of scimitar cats around 36,000 radiocarbon years ago may have been somehow related to arrival of *Arctodus simus* around that same time.[4]

Again, it was a snapshot of remains found around Fairbanks, and Beringia was a complex and giant geographic area with different ecosystems and conditions. For instance, bison and horses made up about 82 percent of the megafauna species in the interior part of Alaska during the late Pleistocene, according to a 1968 paper. Meanwhile a 2013 study found horses were 41 percent of megafauna in Alaska's North Slope at the time, with about 23 percent bison and 16 percent caribou. A Russian paper a year earlier found that 51 percent of the late Pleistocene megafauna in the Siberian portion of Beringia was caribou, with horses making up 25 percent and bison 17 percent.[5]

Aside from the killing fields at La Brea, more giant short-faced bear bones have been found in Alaska than anywhere else in the United States. By the time Ron Richards in Indiana was collecting all the giant short-faced bear data he could find for his 1995 report, there were about a dozen giant short-faced bears listed for Alaska, and a total of twenty-five when combining Alaska and the Yukon. The number has only grown since then.

Although many of the finds have been near Fairbanks, *Arctodus* had made appearances on the far northwestern edge of Alaska too, not far from the Chukchi Sea and the area that would've been laid bare when the land bridge was exposed.

In the summers of 1907 and 1908, an intrepid scientist named L. S. Quackenbush led a small expedition to Eschscholtz Bay, about thirty miles north of the Arctic Circle along the Chukchi. The trip, organized by the American Museum of Natural History, was meant to follow up on reports of mammoths and other prehistoric fossils in the area. He found what he was looking for, plus the remains of a bear, which some have speculated was a giant short-faced bear. Just over a century later, National Park Service archaeologist Jon Hardes was at a fish camp in the same area, helping a local categorize her bone collection, and identified a twenty-five-inch upper arm of an *Arctodus simus*. A guest of the camp had found it sticking out of the beach cliffs near the village of Kotzebue, Hardes told me. There's a photo of him holding it and clearly needing both hands. Around the same time, an oil worker showed up in Hardes's office with a giant short-faced bear mandible that he'd found near the village of Selawik. It looks to measure about a foot long, and at the end there's a massive canine tooth that still looks like it could make easy work of any bit of flesh.[6]

Beringia's huge inhabitants, though, weren't content to remain in one place. After arriving from Siberia in the west, they spread out, filling much of Alaska that wasn't glaciated. There were limits, though. To the east was the advancing and retreating western edge of the Laurentide Ice Sheet, the largest ever known, which sometimes draped over some five million square miles from Nova Scotia and the Far North and east to the Canadian Rockies. Those Beringians who tried to wander south sometimes ran into the Cordilleran Ice Sheet, which often abutted the Laurentide and spread south down into what is today Washington state.

There came a time, around fourteen thousand years ago, when a narrow, ice-free corridor opened up between the two ice sheets. To the south, warmer climates, abundant food, and new territory beckoned. Mammoths, bison, and others pressed into the continent, joining wildlife already living in what is today the Lower 48 states. And at some point—scientists still don't agree when and in what quantities—a bipedal, mostly hairless creature came too.

The more I touched bones and read scientific descriptions, the more I wondered: How many humans ever saw a giant short-faced bear? How many looked into the face of this curious, long-legged marauder and felt the quake of ancestral fear from deep within the trappings of their DNA? It's unknowable, of course, but people and *Arctodus* walked the same continent for thousands of years. The odds are very good that there were more than a few heart-racing encounters.

The scientific record is predictably thin on interactions between people and *Arctodus*. The imagination, though, easily fills the void. A giant short-faced bear on all fours could look a six-foot-man straight in the face, kill with one thrust of its paw, snap bones with a single bite, gash flesh with a single swipe of its claws. They were fast, strong, and well suited to inflict split-second maximum damage. On a good day, primitive weapons like stone-tipped lances would've had limited success piercing thick hides and heavy bones. On a bad day, they would only serve to annoy and anger. There are more than a few modern stories of grizzly bears being shot mid-attack and carrying on nonetheless. *Arctodus* was bigger, stronger, and faced far less-lethal weaponry. It's safe to say that tangling with one of these bears would've been a life-altering—if not *life-ending*—affair.

So fearsome were these bears that it's possible that their mere presence in eastern Beringia, at the far northern reaches of North America, may have delayed the arrival of humans traveling from Siberia into the rest of the continent. That idea has been offered by Valerius Geist, a Canadian biologist with a specialty in the social dynamics of large mammals in North America, and one of the few scientists to fully imagine and articulate the moments of contact between early people and these mighty bears.

"The greatest obstacle of humans in North America must have been the huge carnivores, especially the giant short-faced bear," Geist wrote in a 1993 book.

> No kill by hunters would have been defensible against this beast; no hut would have been safe at night; no human would have been able to outrun this bear; and few trees would have been present or tall enough to climb for safety in the open country where it roamed. The smell of broiling meat and fat would have been irresistible to the bear. Skeletons of giant short-faced bears fill natural cave traps and tar pits, while those of grizzly bears do not. This indicates that the former went heedlessly after tempting prey. Fragile narrow spear blades would have been useless against its 550- to 700-kilogram (1,200 to 1,540 pound) bulk. The only sane response from humans would have been to stay away.[7]

He further noted, "In North America, human occupation began when that of the giant short-faced bear ended" and added, "It appears that large carnivores, plus many ecologically specialized large herbivores, made human life in North America impossible. Only when this fauna collapsed did humans make inroads."[8]

That idea has been disputed by other scientists who say the radiocarbon dating on bones and the timing of the arrival of humans in North America—still an unsettled and hotly debated subject—make clear that *Arctodus* didn't act as a kind of impregnable ursine barrier between Beringia and the rest of the continent. Indeed, there's compelling evidence that people arrived on the continent far earlier than those land wanderers from the north, and strong speculation that some arrived along the West Coast, possibly by boat, and worked their way inland, bears or no bears.

Nonetheless, Geist painted a compelling portrait of the perilous difficulties that early people may have faced in the Far North during the late Pleistocene, especially when it came to the suite of outsized meat-eaters on the landscape: lions, saber-toothed cats, dire wolves, and bears. Humans tend to be fascinated by their own perceived supremacy over the wild, but the odds were stacked against them as they vied for the same prey species.

"Man, as a super-predator, would have faced stiff competition, and could have faced at his kills direct confrontations with at least the largest predators. Since humans have historically such great difficulties dealing with the much smaller, omnivorous brown or grizzly bear, how could they have handled the much larger carnivorous *Arctodus*, particularly in an open landscape without trees to climb?" Geist wrote in a 1989 article called "Did Large Predators Keep Humans Out of North America?"[9]

He noted that killing a bear with arrows or spears tipped with bone or stone would have been "a very difficult task" since they're apt to shatter, chip, or fail to penetrate bone, especially when striking something so large as *Arctodus*. The prospect is made even more daunting given that the bear's heart and lungs were so well hidden and protected. "Stone points do cut very well through soft tissues, as good as iron points or better, but if the projectile is aimed at the heart, it is not at all certain to reach it," Geist wrote.

Even if the weapon pierced the chest of a bear, the wound channel would be so narrow that it'd be unlikely to fully disable the bear, and death might come slowly. Geist noted that soldiers with sixteenth-century Spanish explorer Francisco Vázquez de Coronado plunged a lance into a grizzly, burying it halfway up the shaft, and the bear still managed to catch the rider's horse and maul it before it was finally caught and killed. The same challenges would have faced anyone going up against a giant short-faced bear. "Their weaponry, good for caribou, would not have been of much use," Geist concluded, "but would only enrage the giant bear."

This is something Blaine Schubert has thought a lot about. He told me that one of his "dream fossil sites" would be not one that features a bear killed by a person—though that would be very interesting—but one where a person was killed by a bear. Clearly people and giant short-faced bears overlapped temporally and geographically. "People had to contend with them," Schubert said, "but we haven't really answered the questions about how that went."

So far, there's no evidence that people hunted giant short-faced bears. And the record remains exceedingly scant when it comes to physical indications that early people had anything to do with *A. simus*. One hint, I learned, was lurking inside a squat, one-story building in the rolling plains of West Texas.

20

Lubbock

Long before Buddy Holly, Texas Tech University, the cotton industry, military outposts, Singer's general store, Spanish missions, and even the vast dominion of the Comanches, people have been coming to this low spot in the Southern High Plains looking for a little relief. The area around what is today called the Lubbock Lake Landmark has been inhabited by people for some twelve thousand years—an enviable record for anyone attempting to claim staying power for their old city.

I arrived in Lubbock, Texas, on a warm spring afternoon, showing up a day before my appointments to take it all in. Lubbock Lake is on the northwest edge of town, just off U.S. 84 (also known as Clovis Highway) and right next to a byzantine athletic complex with more than fifty soccer fields and parking lots lined with SUVs and minivans. The site centers on a horseshoe-shaped bend in Yellowhouse Draw, once a tributary of the Brazos River. Today there's a scattering of buildings, but most of the 336-acre Landmark (a unit of the Museum of Texas Tech University) is a protected marshy preserve divided by snaking wooden boardwalks that locals and visitors alike use for contemplative strolls. The first thing I saw from the road, though, was a tall bronze statue of a giant short-faced bear in the middle of a grassy opening. It was like an ambassador from another time, mid-stride on all fours, mouth open and head up like it had just seen something worth investigating. I let out a little "whoop-whoop!" in the car and drove in for a closer look.

I imagined the prehistoric Clovis people might've had a similar reaction, not to the bear, perhaps, but to this place when they arrived after many tired, thirsty miles. It was a slice of green, wet heaven beating in the heart of the dry, rolling South Plains. During a three-thousand-year period during the late Pleistocene and early Holocene, winters here were mostly mild, the summers cool, and water plentiful. Life seemed to spring up at every turn of the head.

Here's what we know this place was like. Some twelve thousand years ago, a slow-moving, sometimes muddy stream meandered along the valley floor full of minnows, gar, catfish, and sunfish. On the marshy shores were migratory geese, muskrats, ducks, rails, water snakes, and bullfrogs. Sedges flowered, softshell turtles buried themselves in the sandy soil, tiger salamanders skittered out of the way, and leopard frogs put out their staccato calls in search of a mate. Groves of hackberry and gromwell played host to turkeys, shrews, and box turtles. In the open grasslands were voles and giant armadillos alongside lumbering mammoths, camels, llamas, bison, peccaries, antelope, and horses. Others passed through too: bears, wolves, and coyotes, as well as ravens, red-winged blackbirds, and vesper sparrows. "Where there is water, there are animals; where there are animals and water, there is man," said one study of the site.[1]

Over the centuries, time and again, families and bands of people returned to camp around the streams and ponds. They ate, they communed, they killed animals like mammoths and bison, they made tools, cut up meat, feasted on smaller prey like box turtles, and, no doubt, enjoyed the plenty that had been provided by such a magical place.

And, at some point, they had at least one run-in with a giant short-faced bear.

When I met Eileen Johnson in the lobby of the Lubbock Lake Landmark visitor center the next morning, the conversation immediately went to the bronze sculpture of the bear and how proud and fascinated Landmark officials are to have a few remains of *Arctodus simus*. "A lot of people can't believe these bears were right here," she said. "But they were."

Johnson, the site's director and an employee since 1972, is quick, warm, and infectiously enthusiastic about the discoveries at Lubbock Lake and the mysteries that remain. The prehistoric bear may not have been a big player in the larger drama that played out here, she said, and certainly is greatly outnumbered in terms of fragments found, but it's significant nonetheless. Giant short-faced bears are featured on signs around the property, and, on the museum's wall-size mural inside, there's one with a piece of flesh hanging from its mouth.

She led me out of the visitor center and into the same grassy spot where I'd stopped the day before to look at the sculpture. Even though the bear is almost six feet tall, Johnson allowed that the statue could've been built bigger to match its real-life size. The sculpture was made in 1993, though, and the science about the bear has come a long way since then. In the same patch of grass was another sculpture, this one of a pampathere, a giant armored creature shaped like a Volkswagen Beetle. It's from the Pleistocene, too, a relative of the armadillo. Its back was worn and smoothed because kids love to mount themselves atop its shell. A rabbit darted in front of us as we walked, apparently unconcerned about the two prehistoric beasts frozen in time a few feet away.

We crossed a small bridge and a gravel road, then went into the Quaternary Research Center, a rectangular concrete building with a few offices and a large open space on one end that operates as the main lab. Several small gray cardboard boxes were already on the table when we arrived, courtesy of one of the researchers there. Johnson and I donned blue latex gloves, and she opened the boxes and began unwrapping the first of six specimens from their tissue-paper cocoons. I got a strange thrill as each piece came out, especially the long canine tooth that almost filled my hand even though it was badly worn on the business end. There was a heft to it, and I wasn't sure if it was the weight itself or the sense I was holding the tooth of an extinct brute that tore through the flesh of other extinct animals more than ten thousand years ago.

"That's it," Johnson said, almost by way of apology for the small handful of *Arctodus* specimens on the table, "that's our whole collection."

Although the tooth was captivating, it was the radius bone (one of the two forearm bones along with the ulna) that harbored the most important story. History is made up of moments, and this artifact, in its obtuse way, told about the moment when the lives of an ancient bear and ancient person intersected at a bend in a river some twelve thousand years ago.

There isn't a lake at Lubbock Lake, but there used to be. Written records of it go back to the mid-1600s, when Spanish explorers put it on their maps as La Punta de Agua, or the Place of Water. The lake and springs provided water for people, horses, and cattle until at least the end of the 1800s. One report recorded up to ten thousand cows watering there in the 1880s and 1890s. As more people moved in, and irrigation farming took hold, more wells were drilled in the area. By the early 1930s, water from the springs no longer reached the surface, and the lake went dry. Locals, unaware that the underground water table had dropped, hatched a plan to reactivate the springs by digging them out and constructing a reservoir. With funds from President Franklin D. Roosevelt's Works Progress Administration, machinery and men with shovels and wheelbarrows began dredging in the spring of 1936. It wasn't long before some boys found a sharpened stone point in one of the dump piles around the reservoir that was later connected with the Folsom people from around ten thousand years ago. The remains of ancient bison, horses, and mammoths were also found.[2]

The springs never did return, and the focus soon switched from digging for water to digging for secrets of the past.

The West Texas Museum conducted its first explorations in 1939 and by the late 1940s, several kill sites had been found, including the charred remains of ancient bison that were about 9,800 years old. Over the following decades, investigators found that Lubbock Lake harbored extraordinarily well-defined layers of sediment, each a band of material in the ground that you can see like layers in a cake. They were neatly stacked one atop the other over thousands of years. Each stratum is a record of life during that period, including evidence of drought or heavy rains, climate, plants, animals, and people. The cultural record spans from the Paleoindian period—of Clovis people from roughly 12,000 to 11,000 years ago and Folsom people from 10,800 to 10,300 years ago—to a time just a few centuries ago.[3]

Lubbock Lake, with its water and lush vegetation, was an irresistible draw for thousands of years. People came and camped and ate and left. No doubt word about this oasis got around, and other clans came and went. Wildlife too. The records of their visits remained on the ground, were covered up by water or sediment or both. The climate changed; the landscape changed and shifted as water pulsed in or retreated. The evidence left behind in those layers of sediment remained largely unseen by modern people until the excavations started. It was a book waiting to be read if only it could be found, opened, and interpreted.

Johnson arrived in 1972, and, not long after, work began focusing on what had once been a gravel bar that seemed to be heavily used by Clovis people. Among layers of sand and clay, they found bones upon bones upon bones. Soon, a story about the place began to emerge. "These records," said one report later, "represent living communities, not simply death assemblages."[4]

Here, at the bottom of a deep valley with a stream running through it, the Clovis people had operated an animal butchering and processing station. And, unlike other Clovis sites, it wasn't just a site that revolved around killing and eating mammoths. Yes, they fed on mammoth, but also bison, camel, pronghorn, horses, giant armadillo, and smaller animals, such as wild turkeys and turtles. They also harvested marrow, divvied up meat, broke bones to use at the site, and even set other bones aside for making tools and weapons later.

"We think it was a fairly sizable processing site along the banks of the small stream that existed at the time. I think a number of events were going on in addition to procuring meat," Johnson told me inside the research building, a hundred yards or so away from the site where the bones were found.

Eventually six giant short-faced bear bones were found there, all probably from the same large male. It's unclear how the bear died, but Johnson doesn't think it was hunted and killed by the people here. Going after a giant short-faced bear would have been a risky endeavor, she said.

"To me, it's a safety thing. Why hunt it when you have all these other animals around?" Johnson said. "It takes an awful lot today to hunt a grizzly, even with modern weapons. It would devastate a family group if you lose one or more of your hunters trying to hunt a mammoth or a bear."

More likely, she said, the Lubbock Lake bear was already dead, and someone decided to scavenge it. Researchers working with a scanning electronic microscope reported evidence of "cut lines" from a stone blade on mammoth and camel bones and on the foot bones of the giant short-faced bear. Others have questioned that interpretation, saying the marks could've come from rolling against rocks in the stream. Johnson and others also looked very closely at the broken forearm bone from the bear that had been found. She has played a key role in developing the scientific interpretation around human modifications of bones, especially in cases when they're fractured. For the bear's forearm, Johnson said, certain kinds of stress and cracks indicate it was broken over a rock.

"This was done while the bone was fresh and in a dynamic manner, a hard, quick impact. That's not the way carnivores break bones," she told me. "This was done by people."

The purpose was revealed when Johnson examined the narrow end of the bone, which she says showed telltale signs of flaking on the edge. The bone, she said, had probably been broken over a rock and then used as a scraper. These kinds of "expediency tools," as she called them, were perfect for stripping meat or working hides. The improvised tools were convenient, available, easy to make, and saved you from lugging around extra stone tools, Johnson said. It seems likely the bear bone was grabbed, snapped in half over a rock, and used either to dress the bear or some other animal alongside the stream. When the work was done, the bear tool was tossed aside and forgotten for thousands of years.

Researchers said the forearm bone was the bear's "first demonstrable association with man." There were two earlier reports, both questionable. At Conkling Cavern in the Organ Mountains of southern New Mexico, there was a report in 1932 that two human skeletons were found "in close proximity" to the remains of a giant short-faced bear and a ground sloth. No additional studies were ever done, though, so the pieces of that puzzle remain unassembled. Farther east in the Guadalupe Mountains, paleontologists in 1935 reported a Clovis point found in a hearth at Burnet Cave with evidence of extinct bison and musk ox. Four vertebrae of *Arctodus* were also found in the cave deposits, but it was unclear how close they were to the hearth and the Clovis points—so it was difficult to determine whether they were truly associated with one another.[5]

While it seems as if the bear bone was one tool at Lubbock Lake, it's curious how few stone tools have been found there—especially considering that New Mexico's Blackwater Draw, the world's most famous Clovis gathering site where thousands of tools have been found, is just over a hundred miles away.

"I think Lubbock Lake was very well known. People came here time and time and time again," Johnson said, before wondering aloud about the connection and differences between Lubbock Lake and Blackwater Draw. "Was that the party place and this the grocery store? We just don't know."

21

Ancient Hunters

On my way to Lubbock from Tucson a few days earlier, I took a detour through the plains of eastern New Mexico, one of those magical kinds of places that seems to exist out of the norms of time and distance. I wanted to see where the Clovis story began, at least in terms of its arrival in modern consciousness.

After a strange night in Roswell in an alien-themed hotel off the highway, I made my way to Eastern New Mexico University in Portales, New Mexico, home to Blackwater Draw Museum, a little brick building attached to a post office on the eastern edge of campus. The museum itself was small but artfully crammed with artifacts and signs that inform with a bit of humor and humanity. I'd heard an *Arctodus* skull was on display. It turned out to be a replica, likely based on one of the La Brea bears, but it still made for a startling impression on a wooden display table in the middle of the museum. It was more than a foot long with a brown sheen and teeth that seemed to smile in greeting like an old friend. There was also a giant short-faced bear featured on a "Predators of the Pleistocene" banner on the back wall, along with a saber-toothed cat.

I asked the college student working at the front desk about any other short-faced bear fossils. That's it, she said, but then added hopefully, "at least for now."

She was kind enough to draw a little map for me to the Black-water Draw site outside of town, taking careful note about where the Walmart is on Highway 70, which was where I'd turn north on Highway 467. She even sent me off with an initialized note that I'd paid to get into the museum and should be allowed to poke around Blackwater Draw without paying again. Her hand-drawn directions were spot-on, and in a few minutes I was rumbling down a pitted dirt road off 467, kicking dust up behind me on my way to one of the most important archaeological sites in the country.

The turnoff was impossible to miss, marked by a green sign and, on a rail just below it, small metal sculptures of a mammoth and a man getting ready to plunge a long spear into his quarry. High drama, indeed. This place, just down the road from the Walmart, is one of the most famous Pleistocene killing grounds in America, a once-lush watering hole where people gathered for thousands of years. They hunted, ate, dug wells, traded stories, formed friendships—and then vanished, but not without leaving behind some clues.

In 1929 a teenager named James Ridgley Whiteman sent the Smithsonian Institution a stone spear point found in a draw between the towns of Clovis and Portales, New Mexico. Researchers came out three years later and conducted their own investigation, eventually finding finger-length spearheads chipped from brittle stone, including some tips buried in the "matted masses of bones of mammoth." The spear points had a distinctive style, thinned at the base ("fluted" is how it's often termed), a telltale mark for the thousands of Clovis points that were eventually found scattered across the country.[1]

The area around Blackwater Draw wasn't just rich with artifacts and bones. The lake and springs at the site had also produced a huge supply of gravel, which, between the 1930s and 1970s, came in handy for building roads in eastern New Mexico. The problem was that the gravel was just below the layers of Clovis remnants, meaning that gravel miners and archaeologists—not to mention local curiosity-seekers—often found themselves digging side-by-side in search of their respective treasures. It was sometimes harmonious, often fractious. Bulldozers ate through dirt in pursuit of profit, scientists howled in protest, artifacts were damaged, historic context was lost, and sore feelings proliferated. Eastern New Mexico University finally bought the site in 1974 and put protections in place.

However fractured and imperfect, there were enough pieces to help paint a picture of what happened at this community spot over hundreds of generations. First of all, some of the leftovers dated back more than eleven thousand years, forcing scientists to recalibrate their notions about how long people had been in North America. Thousands of tools and projectile points were also found, along with bones of ancient horses, bison, saber-toothed cats, and dire wolves. Perhaps most spectacular of all, at least twenty-eight mammoths were found right next to the old lake, many surrounded by knives, spear points, anvil stones, and other tools. These animals were likely hunted by people working in teams; the killing and butchering had been carefully planned and deftly executed with finely made tools honed by hand.[2]

Their technology traveled well. Over the past nine decades, more than 10,000 Clovis points have been found at about 1,500 places in North America. Clovis culture has become synonymous with theories about the first people to arrive on the continent more than 13,000 calendar years ago and with speculation about widespread mammoth hunting. Both of those have been hotly debated, and compelling evidence has emerged that there were likely people in North America before the Clovis, that not everyone arrived via the land bridge between Asia and Alaska, and that the Clovis culture may not have been the fierce mammoth hunters that so many imagined them to be. The discussion continues, in various states of heat and resolution, to this day.[3]

I spent about two hours wandering Blackwater Draw in the middle of the day, seeing barely anyone other than a few employees. The packed dirt path led me in a long circle through the property, slicing through what was once a verdant wetland and now is a scrubby, dry lakebed. At the far corner I stepped into a tall metal building, a little surprised to find the door unlocked and no one else around. Inside was a huge, terraced excavation site. Jumbles of bones jutted out from the gray and orange dirt—bison, horse, mammoth, all manner of creatures once common here were now on full display. Some had been meticulously uncovered by brush-yielding scientists; others still just hinted at what was below the surface. The building had been constructed around the mass of sediment and bones as a way to host visitors, ease the dig, and protect the specimens.

Before I arrived, I had asked around about whether any giant short-faced bears had ever been found at Blackwater Draw. It's hard to imagine such a rich gathering of Pleistocene prey animals—meat, after all—without the biggest meat-eating mammal of all sniffing around. The scientific record there lists about forty mammals, including extinct peccaries, camels, and llamas, along with species we know today, such as foxes, voles, raccoons, and gophers. There were also extinct tortoises and turtles as well as geese, hawks, and garter snakes. Yes, there were black bears. But how could there be no *Arctodus*? Its absence in the fossil record so far doesn't necessarily mean it was never there, of course. It's certainly possible the great bears showed up, filled their bellies, and slipped away only to die somewhere else.[4]

"Are you here, *Arctodus*?" I said to the empty excavation room. "Were you ever here?"

Afterward, back in downtown Portales, I stopped for lunch at the Do Drop In Café, a wood-lined restaurant on the corner. While I was waiting for my turkey sandwich, the only other customer, a bored man in a seed cap who looked to be in his late sixties, started chatting with me and eventually got around to asking what I was doing in town.

"Extinct bears," I said cryptically. I was hot, dusty, and hungry, perhaps less sociable than usual.

"Grizzlies?"

"Giant short-faced bears from the Pleistocene. There might've been one that walked right through here," I responded, absently pointing over my shoulder toward the Clovis site outside of town.

I was half-tempted to tell him that some teeth and toe bones of an *Arctodus* had once been found right here in Roosevelt County, not far from Blackwater Draw, and that a huge arm bone had been uncovered at an Albuquerque gravel pit near the Rio Grande. And did you know they've also been found at a couple caves down in Bernalillo County—a skull in one of them? A few bits of this prehistoric bear were even found back in 1935 in Burnet Cave, over in Eddy County. New Mexico had been a hotbed of short-faced bear discoveries, but I held back to see if the name generated any glimmer of interest.

"Never heard of 'em," he shrugged and let the conversation die on the spot.

Later that day, as I drove south toward Lubbock, I stopped for gas in Littlefield, Texas, and then found out it was the birthplace of Waylon Jennings. The Waylon Jennings museum, I found out, was in the back room of a drive-through liquor store on the corner. Waymore's Package Store, it's called. Who would dare miss that? It was exactly what I hoped it would be—a barely organized mishmash of photos, records, clothes, and other Waylon memorabilia—behind a liquor counter. I had the place to myself until the end of my visit when one of the women working there sold me a small bottle of whiskey and asked where I was from and what brought me through town. I was tempted to revive my spiel about the giant short-faced bear but recalled what had happened in the Do Drop In. "Waylon, of course," I said.

The ancient hunters at the Clovis site were still on my mind during my visit to Lubbock Lake. After Johnson showed me the bear bones in the research building, she walked me back to the main visitor center, chatted a while longer, and then left me alone. I walked back outside, past the big bear sculpture again, over the little bridge, and to the old gravel bar where the butchering site, and the *Arctodus* bones, had been found in the sandy sediment. The spot is sometimes unglamorously referred to in the scientific literature as Feature 1 in Area 2 or simply FA2-1. Today, it's kept secure behind a chain-link fence. The ground is weedy, shaded by a scattering of trees, and there's some standing water in an indentation near the middle. It's a fairly unremarkable view, but I stood there for a long time imagining the thousands of people who had come here over thousands of years. They found the thrill of cool fresh water and the peace of mind that comes with an abundant food supply. Right here, there'd been camels, wolves, pronghorn, mammoths, and horses. Surely more than one giant short-faced bear had wandered through following the scent of freshly butchered meat.[5]

As I left, I paused in the parking lot in the shadow of an eleven-foot bronze sculpture of a mammoth and her baby. She's mid-stride with her head, trunk, and tusks pointing skyward, as if she's leading the way for the young one. The life-size replica is an eye-opener for anyone who's wondered what it be like to stand next to such a beast and ponder the mechanics of delivering a fatal blow. I marveled at the work it must've taken to bring one down, much less gut it, separate flesh from bone, and cart the meat any distance. It's no wonder the impromptu tool made from the bear bone came in handy for those intrepid travelers camped along the water's edge so long ago. I was grateful it had been left behind for us to find and ponder over thousands of years later.

On the drive home from Lubbock, I stopped at New Mexico's White Sands National Park, one of my favorite national parks in the Southwest. It's more than two hundred square miles of glaring white gypsum sand undulating in great waves, some of which are perfect for rifling down atop a plastic snow saucer that they sell at the visitor center. Just inside the entrance of the park's main building there was a special display touting a relatively new discovery. Foot tracks from an ancient lakebed nearby told the story of people more than ten thousand years ago chasing a giant Harlan's ground sloth, a large and awkward Ice Age creature with sharp claws that could be ten feet tall and weigh more than three thousand pounds.[6]

The display included a mold showing the enormous, bean-shaped footprint of the sloth next to a much meeker human print. Scientists speculated that a group of people had been tracking the poor sloth, with possibly one group distracting the animal while others moved in for the kill. They even identified the "flailing circles" of the tracks showing that the sloth had risen on its hind legs to swing its forearms in defense. "The story that we can read from the tracks is that the humans were stalking; following in the footsteps, precisely in the footsteps of the sloth," Matthew Bennett, one the scientists behind the discovery, said in 2018.[7]

Had these same sloth-chasers spent some time at Lubbock Lake, three hundred miles to the east? Had they ever been to Blackwater Draw, perhaps plunged a spear into the rib of a mammoth before wandering on as ever? Did the owners of these footprints ever know the peculiar fright of seeing a giant short-faced bear?

PART 5

Last Stand

22

Endings

About a month after my trip to Lubbock, my wife, daughter, and I were driving through Utah. We'd spent most of the day on backroads and two-lane highways, slipping north through green valleys and cattle ranches at the foot of the Wasatch Mountains and the little roadside towns that always seemed to include a café with pies, a semi-abandoned antique store with its wares outside the front door, and the crisp brick silhouette of an LDS church surrounded by perfectly curated and cut grass.

We were supposed to be making good time on our summer road trip, driving from Arizona to the Pacific Northwest to see family. Bears, though, were still on my mind, so I dragged us off the interstate in the hopes of seeing one of the last haunts of *Arctodus*. We drove north, then, from Kanab in southern Utah up Highway 89 and through Orderville, Panguitch, Circleville, Richfield, Salina, Manti, Ephraim, Spring City, Mount Pleasant. Finally, at Fairview, right at the edge of the Wasatch Mountains, we cut east and up, up into the mountains. "Do you know where you're going?" my wife asked after several miles of twists on the highway and a few snowfields coming into sight. "Yes. Kind of. There's a reservoir up here. On the right, I think. Just a few more miles, I'll bet."

It was the first day of summer, but the aspens were just starting to put out their first buds. As we topped nine thousand feet, the temperature dropped into the high forties. We drove almost twenty miles before reaching Huntington Reservoir. There were two other cars in the gravel lot at the far end of the 136-acre lake. It's stocked with tiger trout, and a few anglers were on the shore readying their rods.

"Was there a bear here?"

"Supposed to be."

On August 8, 1988, a man named Chris Nielson was digging with a backhoe as part of a reconstruction project for the reservoir. Out of the sticky mud he pulled what looked to be a log. A closer examination revealed that it was actually a bone. A very large one. It turned out to be front leg of a fifteen-foot-tall Columbian mammoth. Part of a long, curved tusk was also found. To the crew's credit, work halted and a concerted excavation began. That summer about 90 percent of the mammoth's skeleton was found during a meticulous recovery process. It was in remarkable shape, thanks mostly to the encasement of mud that had hovered around freezing for thousands of years, acting as the perfect refrigerator—and preservation agent—for this ancient mammoth.[1]

"At the excavation it was so fresh that we thought we could smell rotting meat at one place," said David Gillette, who was Utah's state paleontologist at the time.[2]

Based on the wear and tear of its teeth, the big bull mammoth was likely around sixty years old—granddaddy age for an elephant. It was no charmed life on the edge of this receding alpine lake. Nearly all of his bones showed signs of severe and painful disease, mostly arthritis. The partially digested food in his intestinal tract revealed that his last meal was meager and thin, mostly needles and twigs from a fir tree, sedge leaves and seeds. Finally, around thirteen thousand calendar years ago—a point representing "the very end of mammoth existence in America"—he keeled over and died in a mud bog atop this mountain, far from his ancestral home.[3]

Columbian mammoths were typically plains dwellers, so it was unusual to find one in the mountains at nine thousand feet above sea level. (At the time it was the highest mammoth skeleton ever found in North America.) But, when he died, the Pleistocene and the continent's mammoth species were in their twilight as the climate was getting warmer. It's likely the Huntington mammoth was moving upslope in search of cooler climes in the upper reaches of the Wasatch Mountains. But he wasn't alone.

Several projectile points were also found at the dig site, leading to speculation that Paleoindians may have either hunted the mammoth or scavenged it after finding it dead.

Word got out about the mammoth in 1988 and soon locals were sneaking onto the site and digging on their own, even though it was on federal land and they didn't have permission. A crew was called in to guard the area. That's apparently when the remains of a giant short-faced bear were found and whisked away: a single rib and part of its skull, including several teeth. The story was that someone on the night watch duty took the bear parts and stowed them in a refrigerator. They were eventually returned, but the damage was done. Situational context is crucial in paleontological digs, and removing pieces before their location can been closely documented is like ripping pages from a book and trying to understand what they mean. Although the bear bones were recovered and placed safely in a museum the physical context, including exact proximity to the mammoth and the human tools, was lost forever.[4]

Still, there was enough to scientifically piece together some of the story of this giant short-faced bear by Gillette, the Utah paleontologist, and David B. Madsen, both of whom worked at Utah's Division of State History. First of all, it was big, likely in the same ballpark of the giant found in the early 1980s near ancient Lake Bonneville that was estimated to weigh around 1,400 pounds. The Huntington bear also had large teeth, a tall nasal cavity and an exceptionally squished snout. "This individual was distinctly short-faced, an extreme among the short-faced bears," Gillette and Madsen wrote.[5]

And then there was the matter of the projectile points and other lithic tools found nearby. Some of them were similar to points found in western Wyoming from around 9,500 years ago. Others were comparable to those found higher up in the Rocky Mountains of the same vintage, if not a bit older. Could it be that, at this ancient lake, the mammoth, the bear, and the people were existing contemporaneously, each in their own desperate struggle for survival in a changing world? Maybe.

"The presence of these Paleoindian materials suggests, but cannot prove, that humans were contemporary with the Columbian mammoth and the short-faced bear at the Huntington dam site," Gillette and Madsen said.

Utah has long been a hotbed for Ice Age wildlife discoveries. About two miles away from the Huntington Reservoir, there's a site where American mastodons, extinct bison, and extinct horses have been found. About sixty miles north there's another rich Pleistocene find. Silver Creek, as it's known, includes twenty-nine species, like mastodons, ground sloths, dire wolves, saber-toothed cats, camels, horses, and bison. The date there is from about forty thousand years ago. But what happened at Huntington may have been a final dramatic chapter in the final hours of the late Pleistocene. "I'm guessing *Arctodus* was feeding on our poor, dear mammoth," Gillette, the state paleontologist, said at a community meeting a couple months after the discovery. "Perhaps it delivered the final blow."[6]

It's hard to know for sure, but the frozen dead mammoth, so well preserved in the cool boggy ground, might've been one of the last meals of the Huntington bear. Gnaw marks on one of the mammoth's wrist bones show a groove that matches the size and teeth arrangement of the bear. State officials later revised what they think happened: "It's possible that the bear fed on the carcass and died at the same place," they said.[7]

Or it's possible that a different short-faced bear had dined on the mammoth. It's certainly not out of the question. Years later, scientists published a paper after examining mammoth remains found near Saltville, Virginia, with "extreme examples of carnivore gnawing." The mammoth had died, and its carcass was probably partially submerged in water or mud. And then a wolf and another meat-eating scavenger—possibly an American lion or a giant short-faced bear—had come along and gnawed on its heel bones with enough force to be identified by scientists thousands of years later. *Arctodus simus* remains, including parts of its jaw and a fearsome-looking lower canine, had been found nearby. The exact size of the bear is unknown, but there was enough evidence to say it was large and likely quite capable of stealing and defending any carcass coveted by other, smaller scavengers. "In fact its only likely rivals would've been larger members of its own species," said the paper, which was authored by Blaine Schubert and Steven C. Wallace at East Tennessee State University.[8]

Still there was another dimension to the bear at Huntington that had caught my eye. "This individual was one of the last of the Pleistocene megafauna in North America," a state report said in 1996, "perhaps even the last generation."[9]

While the date may be in dispute—Schubert has since calculated that the Huntington bear lived sometime between 12,764 and 13,058 calendar years ago—the thought of a final, lonely bear, whether at Huntington or elsewhere, struck a melancholy note within me.

My wife and daughter stayed near the car to conduct a snowball fight—not many chances for those in Tucson—and I followed the path down from the parking lot, through a little cluster of aspens and into an opening at the base of the reservoir. A pergola had been built along with a series of interpretive signs detailing the discovery of the mammoth, how it was extracted and preserved, and which museums had played a role. I looked at each sign, read every word, searching for any mention of the giant short-faced bear that had been found in the exact same spot. Alas there were none, but that's not what had me a little low.

What struck me was that this may have been the last stop for the giant short-faced bear. Of course it's extremely unlikely, astronomically so, that the exact bear found at Huntington Reservoir was the last one on Earth—what are the odds that the last one would actually be found? And that we could ever make that determination? Still, somewhere the last of its kind dropped dead and when it fell, the great shroud of extinction descended over a species that had inhabited the planet for more than a million years. Strange as it was, I let myself consider the possibility that the final moments for *Arctodus simus* had ticked away at the top of this mountain, the very place I was standing in the chill of the first day of summer.

In recent years, a term has been coined to signify the final individual of a species: an endling. It's an oddly charming term that's poignant and sorrowful and final. I wondered if this Huntington bear was the endling for giant short-faced bears. Had his final meal, and the final meal for his kind, been eaten here? Had he spent years searching in vain for a mate, driven by the indefatigable pursuit of procreation at all costs? Had he wandered into the mountains in a last-ditch attempt to outlast the changing world around him only to find a sickly mammoth on the same lost and doomed path?

And then I wondered why I cared about this particular bear. Surely thousands upon thousands of *A. simus* had perished on the continent, so maybe this one should matter no more than those. But for whatever reason it did. The mind does funny things, and while I stood for a few minutes in the little covered pergola, I ticked through some of the other endlings I knew.

Martha was the name of the last of the passenger pigeons, a species once so populous that when flocks of hundreds of millions flew overhead, day turned to dark. They were mostly gone by the turn of the twentieth century. Martha was born in captivity, spent twenty-nine years at the Cincinnati Zoo, and died in 1914, taking all of the species' genetic and cultural information with her.

Benjamin was the name given to the world's last Tasmanian tiger, a sleek carnivorous marsupial with tiger stripes and a kangaroo pouch. He was captured in the wild and held at Australia's Hobart Zoo under less-than-favorable conditions. He died in 1936, just months after a ban on hunting the species was put into place. Since then, though, there's been considerable debate about whether Ben was actually the last of his kind.

I was always fond of Toughie, the last of the Rabbs' fringe-limbed treefrogs. Originally from Panama, he spent his last years living alone at the Atlanta Botanical Garden, a sort of stately but tragic ambassador of the story of frog extinctions happening around the world. He died in 2016.

And it's hard not to love Lonesome George, the last Pinta Island tortoise, a giant tortoise subspecies, from the Galapagos Islands. He lived to be more than one hundred years old, was never able to breed, and died in 2012, possibly of a heart attack.

It's a mournful record but only a fraction of the extinctions that have happened during our lifetimes. Many went unnoticed, and nearly every endling went unnamed and uncelebrated. What name would we have given this *Arctodus* endling in the Utah mountains if we knew indeed he was the last? Huntington? Wasatch? Björn?

Without a good answer, I walked back up the hill to the car, pausing to take in the long stands of aspens and, at their feet, the snowfields stubbornly hanging on against the season's change, the same way a child clings to a parent's legs when trouble is afoot. I drew in a long breath of the thin mountain air and let it go. Nothing really lasts.

23

What Happened?

In the fall of 2007, a scientific paper with twenty-six authors appeared in the journal *Proceedings of the Natural Academy of Sciences* with a startling assertion: they'd found evidence of an "extraterrestrial impact" from 12,900 calendar years ago that contributed to all of those megafaunal extinctions at the end of the Pleistocene. In short, they posited that one or more "large, low-density ET"—extraterrestrial—objects had exploded over North America ("most likely a comet," they said), triggering massive cooling and new glaciations. First, though, it unleashed fires across thousands of miles, hurricane-force winds, and food shortages that played a key role in dozens of extinctions and important shifts in people trying to make a living on the continent. Part of their argument hinged on what had been found at a place called Murray Springs in southern Arizona.[1]

Intrigued, I visited that spot on a Saturday afternoon. It's about eighty miles southeast of my house in Tucson, just east of the Huachuca Mountains, at the end of a half-mile gravel road that's easy to miss. The creosote bushes were in bloom, a spring rainstorm was kicking up, and desert lizards seemed to skitter beneath my feet at every turn. Otherwise, it was deserted, though that wasn't always the case.

Around thirteen thousand years ago, a small group of nomadic people came to Murray Springs to camp, likely because there was freshwater from the springs, which not only quenched their thirst but proved irresistible to all manner of animals looking for a drink in the arid Southwest. When I was there, standing on the edge of an embankment looking over a dry wash, it was easy to imagine how the ground might've thundered under foot with a passing mammoth and the stir it would've caused for the people there. The remains of two butchered mammoths and eleven bison have been found at the site along with the bones of dire wolves and horses, and hundreds of Clovis artifacts like points, flakes, hide scrapers, knives, and a hearth. It's one of several dozen Clovis sites clustered around the San Pedro River and one of the most well documented in the country.[2]

But lately the discussion has focused less on the bones and tools and more on the curious thin black band of soil found along the walls of the dry wash at Murray Springs and other Clovis sites around the country. According to the scientists in that 2007 paper, the "black mat" essentially drapes over the mammoth and Clovis artifacts—and nothing of the sort appears more recently in the soil record, the dirt that's closer to the surface. It's a dividing line, they say, between the time when those late Pleistocene residents show up in the geologic record, and when they don't. More darkly, it may be a strong signal of catastrophe delivered from outer space.

The scientists, with R. B. Firestone from the U.S. Department of Energy's Lawrence Berkeley National Laboratory as the lead author, provided a detailed chemical analysis of the black layer from Murray Springs and other sites, including Blackwater Draw in New Mexico. Included were a spike in bits called "magnetic microspherules," like tiny flecks of glass, that had been used as evidence for at least eleven other extraterrestrial events around the planet. The paper also described "magnetic grains" along with soot, charcoal, and "glass-like carbon containing nanodiamonds." It all helped paint a portrait of disaster. "The evidence points to an ET event with continent-wide effects, especially biomass burning," the paper said.[3]

Shouldn't the comets, or whatever they were, have left behind a crater? The authors speculated that the fragments—possibly a mile or so across—may have broken up as they barreled toward the Earth and then may have struck the giant Laurentide ice sheet. Once the glaciers melted, any craters that may have existed melted away too, they speculated. Whatever the case, the event would have generated "a devastating, high temperature shock wave" followed by intense winds, toxic ash, and wildfires that would have decimated forests and grasslands, robbing key food sources for herbivores and fatally destabilizing much of the North American food web.

I let those apocalyptic visions wash over me at the site. *Here comes hellfire and damnation. Bye-bye mammoths, adieu to giant ground sloths, farewell to camels and horses, and fare-thee-well to our friends the giant short-faced bears.* It's a hypothesis, anyway.

The black band in the sediment wasn't hard to find when I was there. I spent some time poking around the dry wash, with an eye peeled for rattlesnakes hiding in the shrubs, and spotted it just around a bend where the water once ran. It was at about eye level on one side of the orange and white embankment, maybe a couple inches deep. I flicked a tiny bit with my fingernail and a few dark crumbs fell loose. Certainly to my untrained eye, there was nothing amiss and no hint to me that it was harboring evidence of a life-devouring extraterrestrial event. But what did I know?

The idea isn't particularly outlandish. Some sixty-five million years ago—in one of the worst days on planet Earth—an asteroid some six miles across crashed into what today is known as the Yucatan Peninsula, obliterating about 75 percent of all species living on the planet at the time. And in June 1908 a small asteroid or comet exploded over Siberia's Tunguska River and leveled forests for eight hundred square miles. Outlandish things happen.[4]

But certainly not everyone was convinced that the black band at Murray Springs and the other sites is telling that story. In 2010 a separate group of scientists set out to test the claims by Firestone and others. They sampled at many of the same places that the other researchers had. Their analysis, led by C. Vance Haynes of the University of Arizona, pointed to different explanations, including that some of the material linked to outer space may have simply been some of the copious cosmic dust that falls on the planet each year, and that the burned bits may be more likely explained by hearths used by the Clovis campers. Some of the elements that the 2007 researchers found in the black band at Murray Springs weren't found by the second set of scientists.

"From the data presented here, we find no compelling evidence for a cosmic catastrophe at Murray Springs Clovis site," the scientists said, later adding that their findings didn't necessarily preclude that scenario but simply that their evidence didn't support it.[5]

There was a subsequent response from Firestone and others pushing back against Vance's paper—and the friendly debate has continued ever since.

So what really happened?

What's clear is that some 50,000 years ago, much of the world was populated with very big beasts, fantastic mammals of all stripes that had flourished through the Pliocene and most of the Pleistocene. There were giant deer and mammoth and rhinoceros in Eurasia. South America had car-sized glyptodonts and three-toed litopterns, which looked like hastily assembled grazers made from equal parts horse, camel, and elephant. Australia had hog-sized wombats and the world's biggest marsupials. North America had giant beavers, sloths, mastodons, and saber-toothed cats. And 40,000 years later, they had all vanished. Worldwide, 90 genera of mammals that typically weighed 100 pounds or more were just gone. The extinction event tended to steal the biggest of the big. In North America, south of Alaska, every species that weighed more than a ton had disappeared by the end of the Pleistocene. More than half of the species with an average weight between 32 pounds and 2,000 pounds also came to a bitter end. For those between 10 pounds and 32 pounds, it was 20 percent.[6]

Clearly something happened. What exactly? The question has vexed scientists for more than a century, and it vexes them to this day.

Fossil records are disorganized and sometimes inscrutable things. Often the left-behind remains are stumbled upon while digging in a crop field or exploring a cave or strolling along a sea cliff. If we're lucky, the bones aren't pulled out immediately, and an orga-nized excavation takes place in which scientists can take careful note of the geologic context and conditions. Sometimes, though, the bones are pried out by whoever found them and whisked away to a basement—lost to science evermore—or simply ignored and forgotten. Protected places like Rancho La Brea or Lubbock Lake are a gold mine for paleontologists because they can methodi-cally move strata by strata through the layers of time and derive a sequence of events. But that's not always the case, so reading the fossil record and interpreting past events usually becomes a com-bination of science, knowledge, good fortune, imagination, and the ability to piece together a logical narrative that's supported by whatever fragmented, incomplete evidence has been left behind.

One of the key questions over the extinction of so many mega-faunal species in North America at the end of the Pleistocene was this: Did it really happen in such an all-of-a-sudden way? Argu-ments persisted for years that either a) something catastrophic must've happened to wipe out so many species in such a short period of time or b) actually these large animals winked out slowly over time, perhaps not so much faster than that natural extinction rate that periodically whisks species off nature's stage.

Researchers from George Washington University and the Uni-versity of Wyoming published a paper in 2009 that looked at the timing (and probable timing) of the extinction of saber-toothed cats, giant short-faced bears, mammoths, American lions and other big Ice Age mammals. Of the 35 extinct genera, 16 had a "last appearance date" in the fossil record dating from between 13,800 calendar years ago and 11,400—the very, very end of the Pleistocene. As for the rest, the question became whether they simply died out earlier or if their absence from the fossil record during that time period was the result of sampling error. That is, whether those species were present during that time, and quickly on their way to extinction, but hadn't been found because they were becoming rarer. Using statistical analysis, the researchers determined that most of the extinctions, somewhere between 23 and 31, happened abruptly and around the same time: 13,800 to

11,400 years ago. "Whether or not background extinctions took place," they wrote, "that a catastrophic event or process occurred at the end of the Pleistocene is abundantly clear."[7]

Argument and speculation over what that catastrophic event might've been has run wild for years and has, in a slightly messy way, fallen into handful of rough categories with their own nicknames, some of which are tongue-in-cheek. The "overkill" hypothesis posits that newly arrived people, armed with weapons and evolutionary smarts, killed off all of those mighty beasts in a "blitzkrieg" that swept the continent within a thousand years. "Overchill" backers say that a dramatic climatic shift toward colder conditions— the deep freeze at the end of the Pleistocene called the Younger Dryas—remade the landscapes and provided hostile conditions that larger animals simply weren't able to adapt to fast enough. "Over-ill" raises the possibility of a pandemic that swept the continent after the arrival of the first people. Finally, "over-grill" involves outer-space objects crashing violently into the Earth, or just above it, with cataclysmic consequences, especially for those at the top of the Pleistocene food web.

But it's the "overkill" hypothesis that has spent the most time in the brightest glare of the interrogation spotlight.

Although the rough concept had been around for a long time, it didn't catch fire until the late 1950s and 1960s, when geoscientist Paul S. Martin at the University of Arizona pressed the idea into the public consciousness with a series of papers and discussions. The basic concept was this: when comparing the timing of late Pleistocene extinctions around the world (they didn't all happen at the same time) with the arrival of people, a pattern emerged, what some have called a "deadly syncopation."

"To me the core piece of evidence for human involvement is that when viewed globally, near-time extinctions took place episodically, in a pattern not correlating with climate change or any known factor other than the spread of our species. Extinction followed prehistoric human colonizations," he wrote in his book *Twilight of the Mammoths*. "Simply stated, as humans moved into different parts of the world, many long-established huntable animals died out."[8]

People had several deadly advantages, including stone and wood weapons, the ability to organize, and animals that had never encountered humans before and therefore lacked any instinct to flee, thus making them much easier to kill.

In North America, based on radiocarbon dating from Clovis sites and the presumed extinction dates of some species, the arrival of these hunting people at the very end of the Pleistocene fit neatly with the subsequent disappearance of mammoths, mastodons, great bears, and other megafauna, Martin said. It's not to say that early people killed and butchered every large mammal on the North American landscape but that their actions—a sustained and highly mobile wave of hunting after humans traveled south from Beringia—had both immediate lethal effects and indirect lethal effects as the interconnected web of wildlife broke apart in the aftermath.

"The invading hunters attained their highest population density along a front that swept from Canada to the Gulf of Mexico in 350 years, and on to the tip of South America in roughly 1,000 years," Martin wrote in an earlier paper. "A sharp drop in the human population soon followed as major prey animals declined to extinction."[9]

He was careful about casting blame or judgment on any native groups, saying that his idea was a reflection of *Homo sapiens'* interaction with the natural world: "It is important to remember that the extinctions of near time occurred worldwide. To the extent that responsibility is assigned, it belongs to our species as a whole."

Martin's hypothesis was certainly provocative, arriving at a cultural moment when people were more fully waking up to the worldwide environmental damage that our species is capable of. We had made such a mess of our waters, lands, and air; was it so preposterous to think that our ancestors wouldn't wipe out wildlife as well? The idea also had a grand cinematic scope to it. It was so easy to imagine wave upon wave of people setting foot on a continent, weapons in hand, and rushing headlong into an unsuspecting bestiary.

But could it really be? Did people really arrive in such great numbers and with such absolutely dominating lethal force—the "blitzkrieg"—that eventually every single species weighing more than a ton came to grief, along with many of the other megafauna?

Martin's hypothesis has had its supporters as well as its detractors. Among the central points of criticism are the lack of Clovis sites showing widespread killing, disputes over extinction dates, and, perhaps most important, ever-emerging speculation that people arrived in North America far earlier than previously thought, even before the ice-free corridor opened up between Beringia and the rest of the continent to the south. That could mean people and megafauna coexisted for thousands of years without mass-killing that triggered so many extinctions.

At its most basic level, the narrative of Martin's tsunami-like wave of death traveling across the country, leaving wholesale extinctions in its wake, still leaves many scratching their heads.

"Even allowing for the sake of argument that early paleohunters could have traveled rapidly through most of the physical settings they would have encountered, it is quite another thing to imagine that they could have reduced every megafaunal species along the way to such an extent that in the vast majority of cases there was no survivorship, no recovery *at all*," Ross D. E. MacPhee wrote in his 2018 book *End of the Megafauna*.[10]

If not overkill, then what? Given a lack of plentiful evidence, others have speculated that the sudden drop in temperatures between around 12,900 and 11,500 years ago may have helped spell the end of the big animals in North America. The argument doesn't hinge on physical evidence of emaciated mammoths or other species but more on the idea that, in the absence of another convincing culprit, the sudden cold snap of the Younger Dryas may have done the trick. It wasn't just that the megafauna suffered in the cold, the argument goes, but that it brought about a change in vegetation types that they couldn't ultimately cope with. Large animals need massive amounts of food to stay alive. Combined with their typically slow reproductive rates, the deep freeze and shifts in food sources may have been enough to push them over the edge.

"This does not mean that people played no role in causing the extinctions," Donald Grayson, a University of Washington anthropologist, wrote in 1991 in a paper that pushed against the overkill hypothesis. "A multivariate explanation may yet provide the best account of the extinctions. But no matter what the human role might have been, overkill was not the prime cause of the extinctions. That cause rather clearly lies in the massive climatic change that marks the end of the Pleistocene in North America."[11]

The "over-ill" hypothesis exists in a similar realm. The idea is that human travelers to North America brought with them some sort of disease that resulted in widespread death. But no pathogens have been found in megafaunal bones, no specific disease has been identified, and many remain skeptical that such an outbreak could have such complete and devastating consequences across so many species.

Although the fireball theory proposed by Firestone and others remains an intriguing topic, much of the focus in solving the riddle of the late Pleistocene extinctions still seems to center on Martin's overkill hypothesis, a sudden change in the climate, or some combination of factors.

In 2020 a group of scientists led by Curtin University's Frederik V. Seersholm tried to get at the puzzle with a study of the bones of more than one hundred animal and plant species found in Hall's Cave in central Texas. They, too, noted large-scale changes in the landscape at the end of the Pleistocene that turned many grasslands into woodlands. The loss of those grasslands had a cascading, collapsing effect on vegetarians like horses and camels and a subsequent ripple effect on predators like giant short-faced bears and saber-toothed cats. As the temperatures began to warm again at the onset of the Holocene, nearly all plants and most smaller animals were able to recover. "The faunal diversity, decimated by megafaunal extinctions, did not recover," the authors said. The world simply wasn't the same after the transitional bottleneck between the two epochs. An entire suite of beasts on the continent was just gone.[12]

So why did the large animals perish, never to recover, and the plants and smaller animals survive? What was the key variable? The authors of the 2020 paper returned to Paul Martin's idea. "Hence, these data suggest that human hunting of large animals, likely together with climate at the end of the Pleistocene, led to the extinction of megafauna in North America," they wrote.

Still the broader question remains unsatisfyingly unanswered. "The truth is that both overkill and climate change suffer from plausibility issues concerning their effectiveness as single drivers of extinction," MacPhee wrote in his 2018 book.

> More than anything, it is the magnitude of the losses they supposedly forced during Near Time that strains our credulity. It is possible to imagine how, over a short period of time, small groups of hunters could deleteriously affect a species, or even several species, badly enough to lower standing population sizes. The same outcome could also occur in the case of severe climate change, if the youngest and oldest members of a species were especially hard hit by reductions in, say, food supply. But as a species becomes rarer or less naïve, or as selection increasingly favors the hardier genes of those that remain, these drivers should be progressively less effective. Yet the megafauna, or at least most of them, died. The conundrum persists.[13]

Grayson, in his 2016 book *Giant Sloths and Sabertooth Cats*, said there ultimately may not be a one-size-fits-all explanation for megafauna extinctions, especially when looking at those that happened at different times and places in the late Pleistocene. The answer might be revealed only after the painstaking task of piecing together the stories of the life, death, and times of each species, be they mammoths or giant sloths. In many cases, there just isn't enough data yet to tell those complicated extinction narratives in a conclusive way. "The only way out of this explanatory morass is to tackle the extinct animals one at a time, building local individual species histories," Grayson wrote. "It will take a lot of effort, but until that has been done, we are not likely to see any deeper understanding of these losses than we have now."[14]

My speculation hardly matters, especially given the scientific heavyweights who have put so much of their life's work into solving the riddle. But I have tried to put myself in the place of a giant short-faced bear right at the end of the species' tenure on this planet. If the death of so many great mammals was brought about by hunting, it seems quite unlikely that the bear itself was hunted into extinction. The task would be too hard, too risky, and it hardly seems possible that it could be done so completely. Who really wants to tangle with a giant bear? Rather, the likelihood is that prey species dropped dramatically, competition among meat-eaters stiffened, and eventually there just wasn't food for everyone. *Arctodus* simply couldn't find enough carcasses to scavenge or slow-moving beasts that it could bring down and turn into a meal. Bears are slow to reproduce anyway, and the ranks of *Arctodus* may have just gotten so slim and shadowy over time that it was unable to keep itself going.

But was Martin's idea so outrageous? Humans have singlehandedly wiped out passenger pigeons, dodos, great auks, moas, western black rhinos, Tasmanian tigers, and so many more species. Even grizzly bears, wolves, and bison in the Lower 48 states were driven to the edge of extinction after the arrival of Europeans. Why not the great mammals of the Ice Age?

One question has always nagged at me, though. Not many animals are as adaptable, mobile, and omnivorous as bears. *Arctodus simus* survived in the frigid lands of Beringia, the plains of Wyoming, even the humid far reaches of Florida. Grizzly bears and black bears survived the crucible of the late Pleistocene. Giant short-faced bears couldn't do the same? Not even in a few pockets somewhere?

It seems not. In 2010 Blaine Schubert from East Tennessee State University published a paper looking at radiocarbon dates for giant short-faced bears. The youngest was from a specimen found at

Bonner Springs in Kansas from a bear that may have tipped the scales at more than 1,400 pounds. One earlier date showed it lived around 10,960 calendar ago and another indicated around 12,800 calendar years ago. Despite that discrepancy, the bear almost certainly lived during the transition between the Pleistocene and the Holocene, something few other large animals from the Ice Age were able to accomplish. That itself tells an interesting story. *Arctodus* "may have been one of the last megafaunal species to go extinct in North America," Schubert wrote.[15]

Still, that didn't answer my question about why *Arctodus* vanished. Maybe there simply wasn't enough room on the continent for three bear species. Someone had to go, and it ended up being the biggest of them all. Over the decades there's been scattered speculation about whether the final nail in the coffin for giant short-faced bears was actually the arrival of another big bruin: *Ursus arctos*, the brown bears (which include grizzlies).

In 1974 Björn Kurtén and paleontologist Elaine Anderson announced their analysis of several bear teeth and other bones found in Wyoming's Little Box Elder Cave a few years earlier. Some were from brown bears and others were from *Arctodus simus*, including a bear that appeared to have been quite old when it died and about the size of the smaller versions found in California. It was the first time that the two bears had been associated with one another in a specific place south of Alaska, they said. They theorized that brown bears left Beringia in the late Pleistocene and wandered through the corridor between ice sheets and into the rest of the continent. They didn't speculate whether the two bear species occupied the cave at the same time but said the giant short-faced bear may have been outcompeted by brown bears, which may have been a factor in their extinction.[16]

How could it be that brown bears, generally the smaller of the two bear species, drove giant short-faced bears off the face of the Earth?

Kurtén and Anderson didn't do much to satisfy my curiosity. Scientists Martina L. Steffen and Tara L. Fulton from the Royal BC Museum and Pennsylvania State University, respectively, filled in some of the gaps in 2018. Fossil remains found in a cave on the north end of Vancouver Island proved to be from both brown and short-faced bears. They both lived there as the Pleistocene came to a close and were separated by only three hundred to five hundred years, Steffen and Fulton reported, with the short-faced bears showing up slightly earlier in the fossil record in the cave.

The bears, though, were on Vancouver Island at a tumultuous time. For thousands of years the island was mostly a well-lit, open woods sort of habitat full of towering lodgepole pines and plenty of big, meaty herbivores like mammoths, mastodons, and oxen. A scavenging opportunist like *Arctodus* would have found regular meals of carrion and smaller prey. Conditions began to change rapidly around 13,500 years ago. As temperatures cooled, more dense and dark forests of hemlock and alder sprang up. The vegetation changed, some bigger prey species moved on, and meals became harder to come by for giant short-faced bears, Steffen and Fulton speculated. Although they could eat plants, meat was still an important part of their diet—and it simply became scarcer. In the past, *Arctodus* may have thrived in the more open environment and outcompeted the brown bears, but now the tables were turned. "When closed forests encroached and large meat packages dwindled"—I loved that euphemism—"giant short-faced bears declined and brown bears recurred," they wrote.[17]

Sure, the bigger bear species would have dominated the smaller bears in distinct interactions (say, in a fight over a carcass), but Steffen and Fulton said the evidence suggests that "one reason brown bears persisted where giant short-faced bears died off is because giant short-faced bears may have been less flexible in adapting to new and rapidly changing environments that impacted the availability or quality of food and possibly habitat."

So maybe that was it. Perhaps it just came down to the fact that *Arctodus* was just a little too big and a little too slow to adapt, especially during such roller-coaster changes in the world around them. Similar shifts in vegetation and food supplies happened elsewhere in North America (though it was by no means uniform) at the terminus of the Pleistocene. Brown and black bears were a bit

smaller and more adaptable, and maybe that made all the difference in terms of determining who made it into modern times and
who didn't. Life is a fickle thing, and even the slimmest of margins has consequences.

Studying bears is both "exciting and frustrating," Schubert told
me, and that goes especially for giant short-faced bears because we
still only have a partial portrait of how they lived and died. Part
of the difficulty is that there is no analog today to what life was
like during the North American Pleistocene, when there were vast
numbers of large herbivores and a full complement of big meat-
eaters. "There's not a place you could go today to see all the megafauna and a large bear. It's an extinct ecosystem that we don't have
a match for today. We don't get to see much of that picture, only
glimpses," he said.

That makes it very hard to produce highly informed speculation
about the fate of *Arctodus simus* and how its world may have collapsed around it at the end of the last epoch. In assigning a cause
for their ultimate demise, Schubert thinks it was likely a combination of a changing climate and hunting that had a lethal effect
on food sources. "I can't put a greater weight on one or the other;
both were happening," Schubert said. "If you're a large omnivorous bear relying on megafauna for food, and most items on the
meat menu disappear, then survival as a species becomes increas-
ingly difficult."

He added this: "There are of course other possibilities too, and
those require more speculation. I can imagine that short-faced
bears could be a real nuisance for people, coming around wanting
to steal a meal. If this was the case, then defending food resources
could have resulted in interactions that lead to the death of both
people and bears. I have my fingers crossed that we will get to see
evidence for some of these interactions with future discoveries."

24

De-extinction

For most of human existence, we really didn't know that animals and plants went extinct. What was here had always been here, our ancestors figured, and always would be. But by the 1700s, the study of fossils began gaining ground in some scientific circles. Of particular interest in Europe were bones found in Siberia and in America's Ohio River Valley. At first glance, they appeared to be elephant bones, but there was a problem. Elephants weren't known from those particular areas—and, what's more, the bones upon closer inspection were markedly different from bones of extant elephants. French naturalist and zoologist Georges Cuvier raised the possibility that these animals may have been from a lost world. Sketches of another fossil discovery, what looked to be a giant sloth, added more intriguing evidence to the hypothesis.

"All these facts, consistent among themselves, and not opposed by any report, seem to me to prove the existence of a world previous to ours, destroyed by some kind of catastrophe," Cuvier wrote in 1796. "But what was this primitive earth? What was this nature that was not subject to man's dominion? And what revolution was able to wipe it out, to the point of leaving no trace of it except some half-decomposed bones?"[1]

We know now, through the work of Cuvier and generations of scientists, that extinction has long been part of the world's natural cycles. Conditions change, species adapt, some vanish, and others appear. Evolution occurs in spasmodic fits and starts in a miraculous kind of circus where the acts and the players seem to change minute by minute, hour by hour, eon by eon. Time is spent in the spotlight and then the exit must happen to make room for the next performer. That steady drip of extinctions is also shattered by periodic, colossal planetary die-offs, convulsions of violence that have profound consequences for enormous numbers of species. The fossil record reveals at least five mass extinction events in the Earth's history, those that have wiped out at least 65 percent of species, and scientists say we're in the midst of a sixth great extinction right now. The first was around 443 million years ago in a cataclysm that took a heavy toll on trilobites, corals, and brachiopods. The fifth was the one that, most famously, wiped out the dinosaurs but also sent many, many other life forms to a bitter end. The most gob-smacking event, during the Permian around 225 million years ago, may have wiped out more than 95 percent of marine species. "If these estimates are even reasonably accurate," University of Chicago geologist David Raup once said of the Permian event, "global biology (for higher organisms at least) had an extremely close brush with total destruction." All of the first five major extinctions were likely the result of a calamitous event, like a massive volcanic eruption or a collision with an asteroid or comet. If you survived the first day, you had to contend with thick dust clouds or toxic debris falling from the sky. The sun was blocked, and so plants died along with the animals that depended on them—as well as the animals that depended on eating those plant-eaters. There were devastating chain reactions as vast systems of life collapsed.[2]

In many ways, we should celebrate that we're even here at all. By some estimates there have been thirty billion species on the planet since multicellular life erupted on the planet during the Cambrian explosion of life some 541 million years ago—and roughly 99 percent have come to the ultimate grief. And there's an immutable certainty that comes with it: once a thing is whisked into the oblivion of extinction, it is not coming back. Not the dinosaurs or the dodo or the American lion or the giant short-faced bear. Others may appear in its place but they're not the same, and a cascade of repercussions follows because of the tangled interconnectedness of all living things.[3]

"If the history of life can be viewed as a drama staged on planet Earth, then it can be seen having repeated intermissions, after each of which the cast on stage changes: some characters, previously important, disappear entirely or assume minor roles; others, in the wings, now move to stage front in major roles; new characters sometimes appear, too, producing a constantly shifting, Alice-in-Wonderland effect," Richard Leakey and Roger Lewin wrote about extinction. "Inevitably, fundamental shifts in *dramatis personae* force fundamental changes in the story line. So it is with Earth history."[4]

By the time that the Holocene really kicked into gear about ten thousand years ago, all of those big mammals from the Pleistocene were wiped off the face of North America, and indeed many of them had exited around the world. It isn't just bodies that we lost. Every extinction took with it unique genes, behaviors, memories, relationships, and "services" that each provided for the ecosystem where it lived. In the same way that each person leaves their mark on the world, however big or small, each species leaves an imprint. Think about the pollinating that bees do, or the seed dispersions that birds provide, or the way wolves keep prey populations in check. Take them away and we live in a different place. It's the same for giant sloths, mammoths, sabertoothed cats, and short-faced bears. Their negation reverberates in ways we may never know.

But what if the end wasn't really the end?

For more than a decade, there's been a growing discussion around "de-extinction," the idea that ancient DNA might be used to resurrect extinct animals and plants. It's more than just a *Jurassic Park* fantasy. Science started catching up with science fiction in recent years in genetic labs around world as researchers have pursued several ways to restore lost species, each tricky in its own way. They include cloning (using genetic material to attempt to create a modern copy, though it only works if you have cells from a living being), selective breeding to inject modern species with characteristics of extinct relatives, and editing DNA (a sort of genetic engineering aimed at closely matching an extinct species).

The idea of reanimating extinct species had been kicked around for a long time but found traction in the public consciousness in earnest in 2013 when twenty-five scientists gathered at the National Geographic Society in Washington DC, to discuss the science and ethics of de-extinction. The idea predictably sparked controversy and inspiration in equal measure. Could it mean the return of great, sun-blocking flocks of passenger pigeons? California grizzly bears? Tasmanian tigers? The great auks, whose last two individuals were strangled by Icelandic fishermen in 1844? Perhaps here was a chance to reverse a few mistakes perpetrated on the natural world by humankind.

Often the conversation came around to the prospect of returning gigantic woolly mammoths to the planet. The benefits of resurrecting them extend beyond their individual return and bleed into larger consequences for restoring healthy ecosystems and even tackling planetary problems like global warming, George Church, a professor of genetics at Harvard Medical School, wrote in a 2013 essay for *Scientific American* called "Please Reanimate."[5]

"For instance, 4,000 years ago the tundras of Russia and Canada consisted of a richer grass- and ice-based ecosystem," Church wrote. "Today they are melting, and if that process continues, they could release more greenhouse gas than all the world's forests would if they burned to the ground. A few dozen changes to the genome of a modern elephant—to give it subcutaneous fat, woolly hair and sebaceous glands—might suffice to create a variation that is functionally similar to the mammoth. Returning this keystone species to the tundras could stave off some effects of warming."

Futurist Stewart Brand who, along with his wife, began a nonprofit centered on the "genetic rescue" of endangered and extinct species, helped bring the issue to the national stage. The process, which would take centuries and require overcoming some thorny scientific challenges, would "be a reframing of possibilities as momentous as landing humans on the moon."[6]

"But why do it?" he wrote in a 2014 essay.

What's the point of bringing back some pigeons that have been gone for a century, or some hairy elephants that disappeared four millennia ago? Well, what's the point of protecting unhairy elephants in Africa or over-specialized pandas in China or dangerous polar bears in the Arctic, or any of the endangered species we spend so much money and angst on preserving?

We protect endangered species, conservationists retort (and most of the public agrees), in order to preserve the richest biodiversity we can, to retain creatures that have important ecological roles, or that people love, or as emblems to protect whole endangered ecosystems. We protect them to learn the science to protect them better. We protect them to undo harm that humans have caused. All those apply to bringing back extinct species, plus some.

Unsurprisingly, the idea has its critics. In a joint essay in 2014, famed biologists Paul and Anne Ehrlich at Stanford University said resurrecting extinct species was a "fascinating but dumb idea." Setting aside whether it even could be done, they offered a case against it on several grounds, including a moral stand opposing funding such efforts when so many still-existing species are in grave need of help around the world: "It is much more sensible to put all the limited resources for science and conservation into preventing extinctions, by tackling the causes of demise: habitat destruction, climate disruption, pollution, overharvesting, and so on. Spending millions of dollars trying to de-extinct a few species will not compensate for the thousands of populations and species that have been lost due to human activities, to say nothing of restoring the natural functions of their former habitats."[7]

Another group of scientists in 2020 offered a slight twist on the idea. They posited that water buffalo and feral camels in Australia, feral hogs in the United States, and other large herbivores thriving in places where they didn't appear naturally (even a population of hippos in Colombia, courtesy of drug kingpin Pablo Escobar) may fulfill some of the niches in the wild once filled by the Pleistocene bestiary. The feral hogs rooting around in the soil, for instance, stimulate the growth of additional nutrients for tree growth and more birds, a role once probably played by peccaries. In Australia, grazing by feral camels has lessened fire risk, something that giant marsupials, now extinct, may have done. "We found that, amazingly, the world is more similar to the pre-extinction past when introduced species are included," Erick Lundgren, an ecologist at the University of Technology in Sydney, Australia, and the study's lead author, told the *New York Times*. They acknowledge the incredible damage that nonnative and feral species cause—as well as the fact that these herbivores are operating without the necessary suite of predators to keep them in check—but it's still an interesting observation.[8]

De-extinction, though, still has a tight hold of the imagination and raises all kinds of provocative questions, including about the very nature of species. Would a woolly mammoth even be a woolly mammoth with this kind of genetic rearrangement? Or would it be simply a hirsute elephant?

And could we one day see giant short-faced bears, resurrected and hungry, lumbering around the American West like grizzly bears?

At a lab at the University of California—Santa Cruz, Beth Shapiro and other researchers have been working to sequence and assemble the genome of the giant short-faced bear. They're using bones and skull fragments found in the melting permafrost of the Yukon. The genome will be compared to the genome of its closest living relative, South America's spectacled bear, in search of evidence of natural selection, in particular whether there are genes that speak to differences in diet and habitat.

The research will help piece together the lives and population dynamics of *Arctodus simus* and other bears, but it's not leading toward the return of giant short-faced bears. It just doesn't come up much in conversations about de-extinction. "I imagine that this is because people aren't too keen on the idea of a giant bear," Shapiro told me. "People don't even really like to think about brown bears being reintroduced into parts of the planet from which they have been extirpated."

There's another reason why you won't see replicas of this giant bear. Cloning requires live cells (Dolly the famed sheep clone was the result of tissue taken from a living sheep). No live cells have been found from animals that went extinct in the Pleistocene, so there will be no exact cloning of mammoths, giant short-faced bears, or saber-toothed cats. Instead, the focus is typically on editing the genomes of similar species—like elephants—so that some of the DNA sequences become similar to the sequences of mammoths uncovered by scientists. If that works and a hybrid cell is the result, an embryo could be created and eventually lead to a hybrid species—something that's mostly an elephant, say, but partly a mammoth.

The hope would be that a hybrid creature would carry some of the key mammoth characteristics, perhaps more hair so that it might be better suited for colder climes in places like Siberia or Northern Europe. And if that's the case the most important aspect would come next: the restoration of biological interactions between species that have been lost since the extinctions, the relationships between plants and animals and places that are the foundation of diversity and richness. From an ecological standpoint, it would do no good to simply resurrect extinct species and keep them in a zoo.

Shapiro has been at the forefront of the discussions around de-extinction, especially when it comes to mammoths and passenger pigeons and in seeking clarity around some of the complicated ethical and scientific questions. "Our goal as scientists working in this field is not to create monsters or induce ecological catastrophe but to restore interactions between species and preserve biodiversity," Shapiro wrote in her 2015 book *How to Clone a Mammoth*, where she described herself as an "enthusiastic realist" about de-extinction, saying it is likely to become scientifically possible in the coming years and ethically justified in certain cases.[9]

"If, like elephants, mammoths helped to maintain their own habitat, then bringing mammoths back to life and releasing them into the Arctic may transform the existing tundra into something similar to the steppe tundra of the ice ages," Shapiro wrote. "This might create habitat for living and endangered Arctic species, such as wild horses and saiga antelopes, and other extinct megafauna that might be targets for de-extinction, such as short-faced bears."

Still, she doesn't think there'd be much appetite for resurrecting giant short-faced bears, a species she described to me as "pant-soilingly scary but awesome."

"Personally, I do not see either a route to bringing short-face bears back to life or a need to bring short-face bears back to life," Shapiro told me. "We hardly have enough room for the smaller predators that exist today and should be allowed to have a fighting chance. Yes, they were amazing creatures, and they are one of my favorite extinct animals. However, they are extinct and, as with every other species that has been extinct for thousands of years, they are destined to remain so."

Indeed, the world's surviving large predators are struggling around the globe. Wolves, jaguars, polar bears, grizzly bears—the biggest beasts among us are often the most feared and therefore the most persecuted. Our will to dominate the natural world almost always takes first aim at those that provide the most competition, so we shoot, poison, and trap those posing the most substantial threat. Perversely, these too are the creatures most likely to draw our intense fascination and gaze. The human condition isn't without complications and contradictions.

Besides, even if giant short-faced bears made it back to this world, against all odds, where in the world would you put them?

In 1996 the Russian government donated fifty-five square miles of land in Siberia to support a wild idea: Could the return of large animals help revive a low-nutrient landscape and contribute to the fight against climate change? Could the return of extinct Ice Age animals like woolly mammoths play a role?

Pleistocene Park, as it's now known, hinges on the idea of returning to a time more than ten thousand years ago when a full suite of mega-beasts roamed the land, munching heavily and digesting grasses and other foods, and keeping the vegetation vibrant. Already they've brought in horses, moose, caribou, musk oxen, bison, and other animals. It could also host those animals produced as part of the de-extinction movement, namely mammoths. Nikita Zimov, who took over as the project's director after it was founded by his father, Sergey, has grand—ridiculously grand, beautifully grand—visions for the park, that it will breach its own borders and spread into much of Siberia and into the far reaches of North America and the Canadian Yukon. Restoring this polar region with vast cold-adapted grasslands, something akin to the mammoth steppe that existed during the Pleistocene, will keep the permafrost from melting, thus preventing the release of trapped greenhouse gases that will severely worsen the climate crisis.

"It will be cute to have mammoths running around here," Zimov told *The Atlantic* for a 2017 story that described Pleistocene Park as having a "dreamy, deep-time" atmosphere. "But I'm not doing this for them, or for any other animals. I'm not one of these crazy scientists that just wants to make the world green. I am trying to solve the larger problem of climate change. I'm doing this for humans. I've got three daughters. I'm doing it for them."[10]

The park, of course, will need predators to calibrate the populations of herbivores and keep the landscape from being overgrazed and degraded. You can't replicate past ecosystems without restoring the role-players who made it work. There are already brown bears and wolverines in the area. Zimov has discussed bringing in Siberian tigers, gray wolves, Canadian cougars, and even some version of extinct cave lions from Northern Europe, should the de-extinction movement get that far.

Arctodus simus isn't on that list. For now.

EPILOGUE

To be in bear country is to inhabit another world, one where we are no longer in a position of full control. Step away from the road and hike into the deep backcountry of Yellowstone National Park, for instance, and your senses sharpen knowing there are grizzlies about. As you walk, your eyes search for big moving shapes and your ears become attuned to each rustle in the brush and snapping twig. Of course, the odds of being killed by a bear in Yellowstone are exceedingly low—just slightly worse than being hit by a falling tree or lightning—but it's the *idea* of death-by-bear that grips us so tight. The white-hot terror of an imagined bear attack is easy to access from the primitive parts of our brains: a reminder that we, while clever, still sometimes exist at the pleasure of mightier beasts.[1]

It isn't just fear that causes such a distinct reaction to being around bears. It's also a connection, a recognition that there's something of them in us, and something of us in them. It isn't humanity, exactly, but more of a cue that we are, at the simplest level, both animals traveling through a difficult world and doing the best we can. Look in a bear's face and you might see curiosity, worry, doubt, and even something approaching joy. We can relate to a bear's funny way of scratching its back on a tree or slumping on a log to rest or even a mother bear's impatient herding of her young as she tries to hurry the family from one place to the next. To witness a bear, even from a distance, is to peer through the thin membrane that separates the human world from the animal one, to transcend our modern, too-often-manufactured existence to something more elemental.

Sometimes when I'm out hiking, I imagine that giant short-faced bears survived the Pleistocene. By cunning or dumb luck, they found refuge somewhere to ride out the stranglehold of oblivion more than ten millennia ago, watching from some safe place as the mammoths and the saber-toothed cats and the camels vanished from the scene. And then, once the storm of extinctions passed, they wandered back into their old world to rejoin the ranks of the other bears still living in North America: polar bears, grizzlies, brown and black bears. Pretending they are still here, I conjure what it'd be like to see one coming through the brush at me. The bear would be towering, catlike, and stiff-legged with its stubby nose in the air, ears twitching, interested but a little on edge. Then there'd be a great noise from its snout as it exhaled a great breath and popped its jaw like grizzlies sometimes do. He'd smell for me, and I, locked in place by dread, might instinctively check for his scent on the air. We'd be frozen in time together for a few moments, a breeze rustling his thick coat while the sun catches a lighter ring of fur on his neck and chest. The bear might even feel the need to stand up on its hind legs to get a better assessment of the situation, stretching itself more than ten feet toward the sky, giving me a perfect view of his meaty legs and the paunch at the midsection. If I could will my legs to move (no guarantee there), I'd back away and hope the bear did the same, leaving me to see another day and the bear to find a potentially more nutritious meal somewhere else. If fortune failed me, well, I'd be metabolized into fuel for one of the biggest bears that ever lived.

And what of the rest of world? The one where *Arctodus simus* still lived among us? Surely we'd be awed by them and translate that fascination into human notions. They'd have their likeness turned into stuffed animals, school mascots, and fancy paintings.

They'd be in fairy tales, films, and advertising campaigns. We might travel hundreds of miles to see one in its natural habitat. People would talk about the time they ran into a giant short-faced bear in the wild, with little need to embellish the size of its teeth or the heights it reached. The lore would become part of our personal histories, of our family histories. Maybe the bears would find a home in our spiritual lives, becoming figures of reverence and metaphors in our poetry.

There would be complications, too. Monsters, real or imagined, always say more about us than them. Had *Arctodus* survived, it would have been treated like other big predators: hunted for sport, lured into traps, caged for our enjoyment, poisoned when deemed too threatening, and eventually converted into rugs, coats, and hats or stuffed and mounted for display. Tens of thousands of grizzly bears once lived from coast to coast in the Lower 48, and today they're hemmed into a few isolated pockets in the Intermountain West. It's hard to figure we'd be any more tolerant of a bigger bear that ate more and roamed farther. But *if* we found a way to coexist, we might see our own place in the world a little differently, even if it meant being a bit more frightened by what's just beyond view. "What would an ocean be without a monster lurking in the dark?" filmmaker Werner Herzog said once. "It would be like sleep without dreams."[2]

But no. That world with giant short-faced bears no longer exists. The bones of every *Arctodus simus* that ever lived are buried in the earth, ground into particles, or scattered across museum collections, hidden in dark drawers, stuffed into boxes, or fitted into displays for our safe, pleasurable viewing. The dispossessed remains exist, but the bodies do not. For those left to sift the rubble and divine their story, there is no closure, really. The best we can do is grieve the loss, speculate on the lives they lived, and celebrate what once was.

I remain struck that a bear of such great size and strength looms so small in our culture and consciousness today. No festivals, no mascot, no sense that we walk the same ground as one of the most magnificent bears ever to live. Dinosaurs have been gone sixty-five million years and yet remain the source of undying allure. Meanwhile, *Arctodus* has been gone only around four hundred human generations—certainly in North America people have lived longer with it than without—and barely left a footprint in our modern world. It may be a different story in the natural world. The bears, like all animals before, played some role in shaping neighboring species that survive to this day. Perhaps pronghorns run faster or coyotes dart more daringly because they once shared the landscape with giant short-faced bears. Maybe grizzlies are better scavengers or bighorn sheep have better eyesight. And of what consequence is this bear to us? I wonder what message from our ancestors is hiding in the strands of our DNA, telling us something of *Arctodus*. Perhaps to be wary of its explosive speed or to let it gladly eat its fill instead of trying to scare it off a carcass. Or maybe that it's worth following, from a distance, because its nose is keenly adept at finding places where other, more huntable species, congregate. Surely some of that knowledge resides somewhere inside some of us. At some point, though, that once-essential information may get shed from our genes out of a lack of necessity. What use is it to worry about long-ago ghosts?

I don't often remember my dreams, but one night on the road, around the time I went into Potter Creek Cave in Northern California, I dreamed about a black bear I once saw living in a giant cage next to a gas station in the tiny town of Mitchell, Oregon. In real life, the bear's name was Henry, he weighed eight hundred pounds, and the owners said he'd been rescued from being shot. When I saw him in 2001 as a roadside attraction, he seemed docile, maybe even at peace with his situation, though I had a hard time imagining that was really where he wanted to be. In my dream, I approached the cage and stood face to face with him, only instead of a black bear it was a giant short-faced bear, something akin to the friendly, soft-faced animal that Charles R. Knight had sketched with his pencils back in 1925. We stared at each other, maybe six inches apart, and then the bear let out a long, warm sigh through its loose lips, the kind that might've journeyed across eons, from his ancient world to my contemporary one. As its breath reached my face, I was startled awake, my heart pounding in the hotel room darkness. I was scared and a little giddy, like maybe I'd seen something I wasn't supposed to see, the veil pulled back on a place where things once thought lost were still secretly living.

At least we have our dreams.

ACKNOWLEDGMENTS

I'm a journalist, not a trained scientist, so this book relies heavily on more than a century of exploration and inquiry by scientists intrigued by the notion that giant short-faced bears once roamed these lands. Needless to say (but I will), this book would not be possible without the tenacious scientists who put their time, energy, and expertise into understanding the Pleistocene and these bears specifically. Their work, some of it done long ago and in relative obscurity, lit the way for me to follow. There are too many to mention here, but they show up in the body of this book or are listed in the chapter notes. I hope I did right by them. I'm particularly grateful for those who spent time with me in the field, assisted with arrangements, allowed me to come into their labs and offices, put up with my questions, shared their knowledge selflessly, and pointed me toward further reading or a dark corner I had yet to explore. They have my thanks. That said, any mistakes here are solely my own.

I'd be remiss if I didn't profess gratitude for our nation's libraries, museums, and repositories, and those who staff them. The University of Arizona and Pima County Libraries here in Tucson, Arizona, were invaluable, but so were scores of other institutions around the country that provided a rare paper, photo, book, or other scrap of crucial information. Librarians, archivists, historians, and other diligent record keepers are preserving our collective story for the common good and, as such, deserve our thanks.

My friend and former newspaper editor Tom Tollefson provided invaluable feedback, notes, and encouragement toward the end of this project. I always value his good humor and persistent attempts to keep some of my misguided tendencies in check.

More than anyone, I'm indebted to my wife, Karen, and daughter, Birdie, for their forbearance while I slipped down this bear-shaped rabbit hole over the course of several years. They gamely joined me on a few trips to check out *Arctodus* locales, put up with my mess of papers and books, and indulged my "bears are everywhere" mantra that I'm sure wore thin after a while. I also happen to know they share my love for bears, so that helps. I couldn't ask for better travel partners across time and space.

NOTES

1. Into the Dark

1. William J. Sinclair, *The Exploration of the Potter Creek Cave* (Berkeley: The University Press, 1904), 3.

2. Paul S. Martin, *Twilight of the Mammoths: Ice Age Extinctions and the Rewilding of America* (Berkeley: University of California Press, 2005), 1.

2. Skull

1. Details about the McCloud River hatchery from Joel W. Hedgpeth, *Livingston Stone and Fish Culture in California* (Sacramento: California State Printing Office, 1941).

2. Details about the train wreck from F. E. Raymond, "Livingston Stone, Pioneer Fisheries Scientist: His Career in California," *American Fly Fisher* 16, no. 1 (1990), 18–22.

3. "No church" quote in Livingston Stone, *Report of Operations during 1874 at the United States Salmon-Hatching Establishment on the M'Cloud River, California,* in *United States Commission of Fish and Fisheries, Report of the Commissioner for 1873–74 and 1874–75* (Washington DC: Government Printing Office, 1876), 461.

4. "Fresh green maiden's hair" quote in Stone, *Report of Operations,* 462.

5. "Discovery of a Wonderful Cave," *Sacramento Daily Union,* November 23, 1878, 4.

6. Details about atlatl from R. E. Taylor, *Radiocarbon Dating: An Archaeological Perspective* (Orlando: Academic Press, 1987), 111.

7. *Proceedings of the Academy of Natural Sciences of Philadelphia*, vol. 3 (Philadelphia: Merrihew & Thompson, 1856), 90.

8. Albert E. Sanders, "Additions to the Pleistocene Mammal Faunas of South Carolina, North Carolina, and Georgia," *Transactions of the American Philosophical Society*, vol. 92, pt. 5 (2002), 41.

9. "2,000 pounds" quote from "Huge Cave Bears: When and Why They Disappeared," *Live Science*, November 25, 2008.

10. Bernd Brunner, *Bears: A Brief History* (New Haven: Yale University Press, 2007), 39–40.

11. Björn Kurtén quote from Gary Brown, *The Bear Almanac: A Comprehensive Guide to the Bears of the World* (Guilford CT: Lyons Press, 2009), 11.

12. Joshua Lapp Learn, "Neanderthals Ambushed Cave Bears as They Awoke from Hibernation," *New Scientist*, March 26, 2018.

13. Ian Stirling, *Bears: Majestic Creatures of the Wild* (Emmaus PA: Rodale Press 1993), 18.

14. Brunner, *Bears*, 41.

15. "Our History," California Academy of Science, https://www.calacademy .org/our-history.

16. Henry Fairfield Osborn, *Cope: Master Naturalist* (Princeton NJ: Princeton University Press, 1931), 263.

17. "The Academy of Sciences," *Daily Alta California*, November 4, 1879, 1.

18. E. D. Cope, "The Cave Bear of California," *American Naturalist* 13, no. 12 (December 1879): 791.

3. A Family of Bears

1. Much has been written about the history and lineage of bears, particularly helpful here were Stirling, *Bears*; and Brown, *The Bear Almanac*.

2. James W. Gidley, "A New Species of Bear from the Pleistocene of Florida," *Journal of the Washington Academy of Sciences* 18 (1928): 430.

4. Bone Trove

1. E. D. Cope, "The Californian Cave Bear," *American Naturalist* 25, no. 299 (1891): 997–1001.

2. John C. Merriam, "Recent Cave Exploration in California," *American Anthropologist* 8, no. 2 (April–June 1906): 221–28.

3. William J. Sinclair, "A Preliminary Account of the Exploration of the Potter Creek Cave, Shasta County, California," *Science* 17, no. 432 (May 1, 1903): 708–12.

4. "Greatest Skeleton of a Primeval Bear Ever Found," *San Francisco Call*, September 14, 1902, 8.

5. Sinclair, "A Preliminary Account."

6. "Mammoth Bear Unearthed," *San Francisco Call*, August 13, 1902, 11.

7. "Greatest Skeleton of a Primeval Bear."

8. "Third Arctotherium has been Unearthed," *San Francisco Call*, September 17, 1902.

9. Jens Munthe, "California Speleology, 1901–1908: The State's First Cave Survey," *Journal of Spelean History* 8, no. 2 (April–June 1975): 13–15.

5. Inside

1. Dave Smith, "The Saurian Expedition of 1905 Participants," University of California–Berkeley Museum of Paleontology, https://ucmp.berkeley.edu /about/history/saurexped1905/saurexped1905_bios.php.

6. Into the Pleistocene

1. Carol Jo Rushin, "Interpretive and Paleontologic Values of Natural Trap Cave Bighorn Mountains Wyoming," master's thesis, University of Montana, 1973.

2. Larry D. Martin and B. Miles Gilbert, "Excavations at Natural Trap Cave," *Transactions of the Nebraska Academy of Sciences and Affiliated Societies* (1978): 107–16.

3. Xiaoming Wang and Larry D. Martin, "Natural Trap Cave," *National Geographic Research and Exploration* 9, no. 4 (1993): 422–35.

4. Julie A. Meachen, Alexandria L. Brannick, and Trent J. Fry, "Extinct Beringian Wolf Morphotype Found in Continental U.S. Has Implications for Wolf Migration and Evolution," *Ecology and Evolution* 6, no. 10 (2016): 1–9.

5. A. R. Wallace, *The Geographical Distribution of Animals, with a Study of the Relations of Living and Extinct Faunas as Elucidating the Past Changes of the Earth's Surface*, vol. 1 (New York: Harper and Brothers, 1876), 150.

6. Ross D. E. MacPhee, *End of the Megafauna: The Fate of the World's Hugest, Fiercest, and Strangest Animals* (New York: W. W. Norton, 2019), 28.

7. Richard Leakey and Roger Lewin, *The Sixth Extinction: Patterns of Life and the Future of Humankind* (New York: Anchor, 1995), 71.

8. Björn Kurtén, *The Age of Mammals* (London: Weidenfeld and Nicholson, 1971), 186–93.

9. MacPhee, *End of the Megafauna*, 195.

7. Bears in Proximity

1. Mike Stark, "Apple-Fed Grizzly Killed," *Billings Gazette*, April 3, 2003.

2. Chuck Neal, *Grizzlies in the Mist* (Moose WY: Homestead, 2003), 148.

3. Neal, *Grizzlies in the Mist*, 20.

8. Bears Are Everywhere

1. "The Lives They Lived: Pedals the Bear," *New York Times*, December 21, 2016.

2. Brunner, *Bears*, 7.

3. List of college sports team nicknames, https://en.wikipedia.org/wiki /List_of_college_sports_team_nicknames#B.

4. Brown, *The Bear Almanac*, 87.

5. Brown, *The Bear Almanac*, 254.

6. Brunner, *Bears*, 25.

7. Pliny, *Natural History* (Cambridge MA: Harvard University Press, 1940). "The bear's (breath) is pestilential" is from book 11, 607. "The bear's weakest part" is from book 8, 93.

8. Brown, *The Bear Almanac*, 44.

9. Landon Y. Jones, *The Essential Lewis and Clark* (New York: First Ecco, 2002), 39.

10. As quoted in Brown, *The Bear Almanac*, 183.

11. Brown, *The Bear Almanac*, 183.

12. Brown, *The Bear Almanac*, 183.

13. John Muir, *The Wilderness World of John Muir* (New York: Houghton, Mifflin, 2001), 313.

14. George Laycock, *The Wild Bears* (New York: Outdoor Life, 1986).

15. "Tribal Witnesses Emphasize Spiritual and Cultural Significance of Grizzly Bears," *Indian Country Today*, May 17, 2019.

16. David Rockwell, *Giving Voice to Bear: North American Indian Myths, Rituals, and Images of the Bear* (Niwot CO: Roberts Rinehart, 1991), 1–2.

17. Alfred Irving Hallowell, "Bear Ceremonialism in the Northern Hemisphere," *American Anthropologist* 28, no. 1 (1926): 22.

9. La Brea

1. Chester Stock, *Rancho La Brea: A Record of Pleistocene Life in California*, rev. John M. Harris (Los Angeles: Natural History Museum of Los Angeles, 2001).

2. La Brea Tar Pits and Museum, tarpits.org.

10. Hiding in the Muck

1. "LAPD Makes Arrest in Baldwin Hills 'Rolls Royce' Killing," *Los Angeles Times*, January 17, 2012.

2. Additional details from "Intradepartmental Correspondence: Request for Payment of Reward Office on City Council File no. 11-0010-592," Los An-

geles Police Department, March 13, 2014, http://www.lapdpolicecom.lacity
.org/040814/bpc_14–0083.pdf.

3. Ben M. Waggoner, "The La Brea Tar Pits, Los Angeles," University of California–Berkeley Museum of Paleontology, 1996, and updated in 2011, https://ucmp.berkeley.edu/quaternary/labrea.php.

4. Stock, *Rancho La Brea*, 2.

5. Stock, *Rancho La Brea*, 2.

6. "La Brea Tarpits History," La Brea Tar Pits and Museum, https://tarpits .org/la-brea-tar-pits-history.

7. John C. Merriam, *The Fauna of Rancho La Brea* (Berkeley: The University Press, 1911), 202.

8. *Proceedings of the Boston Society of Natural History, Vol. XVIII, 1875–1876* (Boston: Printed for the Society, 1877), 185–86.

9. W. W. Orcutt, "Early Oil Development in California," American Association of Petroleum Geologists Bulletin, 8, no. 1 (1924), 61–72.

10. John C. Merriam, "Death Trap of the Ages," *Sunset* 21, no. 6 (October 1908): 465–75.

11. "Death Trap of the Ages."

12. Stock, *Rancho La Brea*, 15–16.

13. Björn Kurtén, *Before the Indians* (New York: Columbia University Press, 1988), 103.

14. Stock, *Rancho La Brea*, 14.

15. Stock, *Rancho La Brea*, 4.

11. Teeth and Bones

1. "Dental Caries in the fossil record: a window to the evolution of dietary plasticity in an extinct bear," Borja Figueirido, Alejandro Perez-Ramos, Blaine W. Schubert, Francisco Serrano, Aisling B. Farrell, Francisco J. Pastor, Aline A. Neves, and Alejandro Romero, *Scientific Reports*, December 19, 2017.

2. "Pit 91," La Brea Tar Pits and Museum, https://tarpits.org/experience -tar-pits/pit-91.

3. Stock, *Rancho La Brea*.

4. "Last credible sighting" quote is from the California Grizzly Research Network, https://www.calgrizzly.com/home-1.

5. Kevin Ferguson, "Monarch: The Sad, Amazing Story of the Bear on California's State Flag," Southern California Public Radio, October 24, 2012.

6. Peter Fimrite, "CA Grizzly Bear Monarch: A Symbol of Suffering," *San Francisco Chronicle*, May 3, 2011.

12. A Surge of Discovery

1. Lawrence Lambe, "On Arctotherium from the Pleistocene of Yukon," *Ottawa Naturalist* 25, no. 2 (May 1911): 21–26.

2. L. David Carter, Thomas D. Hamilton, and John P. Galloway, eds., *Late Cenozoic History of the Interior Basins of Alaska and the Yukon*, U.S. Geological Survey circular 1026 (1989), 96.

3. O. A. Peterson, "The Fossils of Frankstown Cave, Blair County, Pennsylvania," *Annals of the Carnegie Museum* 26, no. 1 (July 1925): 249.

4. Ralph W. Stone, *Pennsylvania Caves* (Harrisburg: Pennsylvania Geological Survey Bulletin, 1930), 54.

5. Oliver P. Hay, *The Pleistocene of the Middle Region of North America and Its Vertebrated Animals* (Washington DC: Carnegie Institute, 1924), 303.

6. W. D. Matthew, *Contributions to the Snake Creek Fauna, with Notes upon the Pleistocene of Western Nebraska*, Bulletin of the American Museum of Natural History 38, no. 7 (April 18, 1918): 183–229.

7. John R. Schultz, "A Late Quaternary Mammal Fauna from the Tar Seeps of McKittrick, California," PhD thesis, California Institute of Technology, 1937.

8. Kena Fox-Dobbs, Robert G. Dundas, Robin B. Trayler, and Patricia A. Holroyd, "Paleoecological Implications of New Megafaunal C14 Ages from the McKittrick Tar Seeps, California," *Journal of Vertebrate Paleontology* 34, no. 1 (2014): 220–23.

9. George C. Rinker, *Tremarctotherium from the Pleistocene of Meade County, Kansas* (Ann Arbor: University of Michigan Press, 1949), 107.

10. Ronald L. Richards, C. S. Churcher, and William D. Turnbull, "Distribution and Size Variation in North American Short-Faced Bears, Arctodus Simus," in *Palaeoecology and Palaeoenvironments of Late Cenozoic Mammals: Tributes to the Career of C.S. (Rufus) Churcher* (Toronto: University of Toronto Press, 1996).

11. Björn Kurtén, "Pleistocene Bears of North America: 2. Genus Arctodus, Short-Faced Bears," *Acta Zoologica Fennica* 117 (March 1967): 1–60.

12. "Arctodus Pristinus," Florida Museum, https://www.florida museum.ufl.edu/florida-vertebrate-fossils/species/arctodus-pristinus/.

13. What Happened in Fulton County

1. "About 35 pounds a day" quote from Jon Hardes, "Ice Age Mammal Bones of Northwest Alaska (#2)," National Park Service, October 21, 2013, https://www.nps.gov/kova/blogs/ice-age-mammal-bones-of-northwest-alaska-2.htm.

2. "The Monster at Manitou," Cass County Historical Society, https://www.casshistory.net/manitau.html.

3. Tim Swartz, "An Unnatural History of Indiana: Indiana's Lake Monsters," *Strange Magazine* 21, http://www.strangemag.com/strangemag/strange21/unnaturalindiana/unnaturalindiana5lakemonst.html.

4. Shirley Willard, "Rochester's Giant Short-Faced Bear," *Outdoor Indiana* 70, no. 1 (January/February 2005): 35.

5. Ronald L. Richards and William D. Turnbull, "Giant Short-Faced Bear (Arctodus simus yukonensis) Remains from Fulton County, Northern Indiana," *Fieldiana* 30 (April 28, 1995): 1–34.

6. Associated Press, "Rochester Bear Syphilis Victim?" *Logansport Pharos Tribune*, September 3, 1987.

7. "Bare Big Bear Bones," *South Bend Tribune*, February 22, 1968.

14. Real Monsters

1. United Nations, "Global Assessment Report on Biodiversity and Ecosystem Services," Intergovernmental Science-Policy Platform on Biodiversity and Ecosystem Services, 2019, https://www.un.org/sustainabledevelopment/blog/2019/05/nature-decline-unprecedented-report/.

2. "The Current Mass Extinction," PBS *Evolution Library*, https://www.pbs.org/wgbh/evolution/library/03/2/l_032_04.html.

3. Blaine W. Schubert, "Late Quaternary Chronology and Extinction of North American Giant Short-Faced Bears (*Arctodus simus*)," *Quaternary International* 217, no. 1 (2010): 188–94.

4. "Some of New Bern's famous bear statues are floating away in Florence flooding, city says," Abbie Bennett, *The News & Observer*, September 14, 2018.

15. A Hoosier's Search

1. Ronald L. Richards, "Gettin' down to Bear Bones," *Outdoor Indiana*, July–August 1983, 33.

2. Richards and Turnbull, "Giant Short-Faced Bear (Arctodus simus yukonensis) Remains."

3. Richards et al., "Distribution and Size Variation."

4. "New Life for Old Bones," Richard G. Biever, *Indiana Connection*, February 21, 2018.

17. Fitful Arrivals

1. Blaine W. Schubert and James E. Kaufmann, "A Partial Short-Faced Bear Skeleton from an Ozark Cave With Comments on the Paleobiology

of the Species," *Journal of Cave and Karst Studies* 65, no. 2 (August 2003): 101–10.

2. Blaine W. Schubert, "Discovery of an Ancient Giant in an Ozark Cave," *The Living Museum* 63, no. 1 (spring 2001): 10–13.

3. Bob McEowen, "A Treasure Below: The Startling Discovery of an Ice Age Cave Rewrites Science," *Rural Missouri*, January 2008, http://www.ruralmissouri.coop/08pages/08JanRiverbluffCave.html.

4. Scott Harvey, "Discovering Our Past: New Film Looks at Ozarks' Cave, Other American Landmarks," KSMU, August 13, 2014, https://www.ksmu.org/post/discovering-our-past-new-film-looks-ozarks-cave-other-american-landmarks.

5. Steven D. Emslie and Nicholas J. Czaplewski, "A New Record of Giant Short-Faced Bear, Arctodus Simus, from Western North America with a Re-evaluation of Its Paleobiology," Natural History Museum of Los Angeles County, *Contributions in Science*, no. 371 (1985): 1–12.

6. Michael E. Nelson and James H. Madsen, "A Giant Short-Faced Bear (Arctodus Simus) from the Pleistocene of Northern Utah," *Transactions of the Kansas Academy of Sciences* 86, no. 1 (1983): 1–9.

7. Jay Van Tassell, John Rinehart, and Laura Mahrt, "Late Pleistocene Airport Lane Fossil Site, La Grande, Northeast Oregon," *Oregon Geology* 70, no. 1 (summer 2014): 3–13.

8. Martina L. Steffen and C. R. Harrington, "Giant Short-Faced Bear (Arctodus Simus) from Late Wisconsinan Deposits at Cowichan Head, Vancouver Island, British Columbia," *Canadian Journal of Earth Sciences* 47, no. 8 (July 2010): 1029–36.

9. "Longest Polar Bear Swim Recorded—426 Miles Straight," *National Geographic*, July 2011, https://www.nationalgeographic.com/news/2011/7/110720-polar-bears-global-warming-sea-ice-science-environment/.

10. Blaine W. Schubert and Richard C. Hulbert, "Giant Short-Faced Bears (Arctodus Simus) in Pleistocene Florida, USA, a Substantial Range Extension," *Journal of Paleontology* 84, no. 1 (2010): 79–87.

18. Imagining *Arctodus*

1. Kurtén, "Pleistocene Bears of North America: 2."

2. Cope, "The Californian Cave Bear," 997–1001.

3. "Greatest Skeleton of a Primeval Bear," 8.

4. Emslie and Czaplewski, "A New Record of Giant Short-Faced Bear."

5. Paul Edward Matheus, "Paleoecology and Ecomorphology of the Giant Short-Faced Bear in Eastern Beringia," thesis, University of Alaska–Fairbanks, December 1997.

6. Grizzly bear speeds from Scott Weybright, "How Do These Big, Boxy Creatures with Flat Feet Run So Fast?," Washington State University, College of Agriculture, Human, and Natural Resources Sciences, June 24, 2016, https://cahnrs.wsu.edu/blog/2016/06/how-do-these-big-boxy-creatures -with-flat-feet-run-so-fast/.

7. David Mattson, "Diet and Morphology of Extant and Recently Extinct Northern Bears," *Ursus* 10 (1998): 479–96.

8. Gennady Baryshnikov, Larry D. Agenbroad, and Jim I. Mead, "Carnivores from the Mammoth Site, Hot Springs, South Dakota," chapter 6 in *The Mammoth Hot Springs Site: A Decade of Field and Laboratory Research in Paleontology, Geology, and Paleoecology* (Hot Springs SD: Mammoth Hot Springs Site, 1994).

9. Boris Sorkin, "Ecomorphology of the giant short-faced bears Agriotherium and Arctodus," *Historical Biology* 18, no. 1 (2006): 1–20.

10. Borja Figueirido, Juan Antonio Perez-Claros, Vanessa Torregrosa, Alberto Martin-Serra, and Paul Palmqvist, "Demythologizing *Arctodus Simus*, the 'Short-Faced' Long-Legged and Predaceous Bear That Never Was," *Journal of Vertebrate Paleontology* 30, no. 1 (2010): 262–75.

11. Shelly L. Donohue, Larisa R. DeSantis, and Blaine W. Schubert "Was the Giant Short-Faced Bear a Hyper-Scavenger? A New Approach to the Dietary Study of Ursids Using Dental Microwear Textures," *PLOS One*, October 30, 2013.

19. The Great and Far North

1. C. S. Churcher, A. V. Morgan, and L. D. Carter, "Arctodus Simus from the Alaskan Arctic Slope," *Canadian Journal of Earth Sciences* 30, no. 5 (1993): 1007–13.

2. Taylor B. Young and Joseph Little, "The Economic Contribution of Bear Viewing to Southcentral Alaska," University of Alaska–Fairbanks, School of Management, prepared for Cook Inletkeeper, May 2019.

3. MacPhee, *End of the Megafauna*, 28.

4. Kena Fox-Dobbs, Jennifer A. Leonard, and Paul L. Koch, "Pleistocene Megafauna from Eastern Beringia: Paleontological and Paleoenvironmental Interpretations of Stable Carbon and Nitrogen Isotope and Radiocarbon Records," *Palaeogeography, Palaeoclimatology and Palaeoecology* 261 (2008): 30–46.

5. Pamela Groves, "Pleistocene Megafauna in Beringia," *Alaska Park Science* 17, no. 1 (2019): 25–33.

6. L. S. Quackenbush, "Notes on Alaskan Mammoth Expeditions of 1907 and 1908," *Bulletin of the American Museum of Natural History* 26, no. 9 (1909): 87–130.

7. Geist quoted in Stirling, *Bears*, 25.

8. Stirling, *Bears*, 18.

9. Valerius Geist, "Did Large Predators Keep Humans Out of North America?," in *The Walking Larder: Patterns of Domestication, Pastoralism, and Predation*, edited by Juliet Clutton-Brock (New York: Routledge Library Editions, Archaeology, 1989), 283.

20. Lubbock

1. Eileen Johnson, ed., *Lubbock Lake: Late Quaternary Studies on the South High Plains* (College Station: Texas A&M University Press, 1987), 4.

2. Johnson, *Lubbock Lake*, 6.

3. "Archaeological Context," Lubbock Lake Landmark, http://www.depts .ttu.edu/museumttu/lll/Context.html.

4. Johnson, *Lubbock Lake*, 160.

5. Johnson, *Lubbock Lake*, 88.

21. Ancient Hunters

1. Charles C. Mann, "The Clovis Point and the Discovery of America's First Culture," *Smithsonian Magazine*, November 2013.

2. Blackwater Draw Site and Museum, Eastern New Mexico State University.

3. Michael R. Waters, Thomas W. Stafford Jr., and David L. Carlson, "The age of Clovis—13,500 to 12,750 cal yr B.P.," *Science Advances*, October 21, 2020.

4. "Blackwater Draw Fauna," list from the University of Texas at El Paso, https://www.utep.edu/leb/pleistnm/sites/blackwaterlocl.htm.

5. Vance Terrell Holliday, "Cultural Chronology of the Lubbock Lake Site," thesis, Texas Tech University, 1977.

6. "Harlan's Ground Sloth, *Paramylodon Harlani*," Explore the Ice Age Midwest, Illinois State Museum.

7. Reuters and Cheyenne Macdonald, "The Ancient Footprints That Reveal How Humans Stalked Giant Seven-Foot Sloths 11,000 Years Ago," *Daily Mail*, April 26, 2018.

22. Endings

1. David D. Gillette and David B. Madsen, "The Columbian Mammoth, Mammuthus Columbi, from the Wasatch Mountains of Central Utah," *Journal of Paleontology* 67, no. 4 (July 1993): 669–80.

2. Gillette quote from official signs at the Huntington mammoth discovery site in Utah.

3. The 11,200 date is from Gary Haynes, "The Catastrophic Extinction of North American Mammoths and Mastodonts," *World Archaeology* 33, no. 3 (February 2002): 391–416.

4. Associated Press, "Bones Provide Understanding of Mammoth in Utah Canyon," *Times-News* (Twin Falls, Idaho), November 21, 1988.

5. David Gillette and David Madsen, "The Short-Faced Bear *Arctodus Simus* from the Late Quaternary in the Wasatch Mountains of Central Utah," *Journal of Vertebrate Paleontology* 12, no. 1 (March 1992): 107–12.

6. Associated Press, "Bones Provide Understanding of Mammoth."

7. David D. Gillette, "Utah's Wildlife in the Ice Age," Utah Geological Survey, *Survey Notes* 28, no. 3 (May 1996): 5–8.

8. Blaine W. Schubert and Steven C. Wallace, "Late Pleistocene Giant Short-Faced Bears, Mammoths, and Large Carcass Scavenging in the Saltville Valley of Virginia, USA," *Boreas* 38, no. 3 (2009): 482–92.

9. Gillette, "Utah's Wildlife in the Ice Age."

23. What Happened?

1. R. B. Firestone, A West, J. P. Kennett, L. Becker, T. E. Bunch, Z. S. Revay, P. H. Schultz, T. Belgya, D. J. Kennett, J. M. Erlandson, O. J. Dickenson, A. C. Goodyear, R. S. Harris, G. A. Howard, J. B. Kloosterman, P. Lechler, P. A. Mayewski, J. Montgomery, R. Poreda, T. Darrah, S. S. Que Hee, A. R. Smith, A. Stich, W. Topping, J. H. Wittke, and W. S. Wolbach, "Evidence for an Extraterrestrial Impact 12,900 Years Ago That Contributed to the Megafaunal Extinctions and the Younger Dryas Cooling," PNAS 104, no. 41 (October 9, 2007): 16016–21.

2. C. Vance Haynes Jr., J. Boernerb, K. Domanikc, D. Laurettac, J. Ballengerd, and J. Gorevac, "The Murray Springs Clovis Site, Pleistocene Extinction, and the Question of Extraterrestrial Impact," PNAS 107, no. 9 (March 2, 2010): 4010–15.

3. Firestone et al., "Evidence for an Extraterrestrial Impact."

4. William K. Hartmann, "The Impact That Wiped Out the Dinosaurs," Planetary Science Institute, https://www.psi.edu/epo/ktimpact/ktimpact.html.

5. Haynes et al., "The Murray Springs Clovis Site."

6. Paul L. Koch and Anthony D. Barnosky, "Late Quaternary Extinctions: The State of the Debate," *Annual Review of Ecology, Evolution and Systematics* 37 (2006): 215–50.

7. J. Tyler Faith and Todd A. Surovell, "Synchronous Extinction of North America's Pleistocene Mammals," PNAS 106, no. 109 (December 8, 2009): 20641–45.

8. Martin, *Twilight of the Mammoths*, 51.

9. Paul S. Martin, "The Discovery of America," *Science*, New Series 179, no. 4077 (March 9, 1973): 969–74.

10. MacPhee, *End of the Megafauna*, 151.

11. Donald K. Grayson, "Late Pleistocene Mammalian Extinctions in North America: Taxonomy, Chronology and Explanations," *Journal of World Prehistory* 5, no. 3 (1991): 193–231.

12. Frederik V. Seersholm, Daniel J. Werndly, Alicia Grealy, Taryn Johnson, Erin M. Kennan Early, Ernest L. Lundelius Jr., Barbara Winsborough, Grayal Earle Farr, Rickard Toomey, Anders J. Hansen, Beth Shapiro, Michael R. Waters, Gregory McDonald, Anna Linderholm, Thomas W. Stafford Jr., and Michael Bunce, "Rapid Range Shifts and Megafaunal Extinctions Associated with Late Pleistocene Climate Change," *Nature Communications* 11, no. 1 (June 2020): 1–10.

13. MacPhee, *End of the Megafauna*, 183.

14. Donald K. Grayson, *Giant Sloths and Sabertooth Cats: Extinct Mammals and the Archaeology of the Ice Age Great Basin*, (Salt Lake City: University of Utah Press, 2016), 294.

15. Schubert, "Late Quaternary Chronology and Extinction."

16. Björn Kurtén and Elaine Anderson, "Association of *Ursus Arctos* and *Arctodus Simus* in the Late Pleistocene of Wyoming," *Breviora*, no. 426 (November 27, 1974): 1–6.

17. Martina L. Steffen and Tara L. Fulton, "On the Association of Giant Short-Faced Bear (*Arctodus Simus*) and Brown Bear (*Ursus Arctos*) in Late Pleistocene North America," *Geobios* 51, no. 1 (2018): 61–74.

24. De-extinction

1. Martin J. S. Rudwick, *Georges Cuvier, Fossil Bones, and Geological Catastrophes* (Chicago: University of Chicago Press, 1997), 24.

2. Leakey and Lewin, *The Sixth Extinction*, 44.

3. Leakey and Lewin, *The Sixth Extinction*, 39.

4. Leakey and Lewin, *The Sixth Extinction*, 46.

5. George Church, "Please Reanimate" or "De-extinction Is a Good Idea," *Scientific American* 309, no. 3 (September 1, 2013): 12.

6. Stewart Brand, "De-extinction Debate: Should We Bring Back the Woolly Mammoth?" *Yale360*, January 13, 2014.

7. Paul Ehrlich and Anne H. Ehrlich, "The Case against De-extinction: It's a Fascinating but Dumb Idea," *Yale360*, January 13, 2014.

8. Asher Elbein, "Pablo Escobar's Hippos Fill a Hole Left Since Ice Age Extinctions," *New York Times*, March 26, 2020.

9. Beth Shapiro, *How to Clone a Mammoth: The Science of De-extinction* (Princeton NJ: Princeton University Press, 2015), prologue and 15.

10. Ross Andersen, "Pleistocene Park," *The Atlantic*, April 2017.

Epilogue

1. Odds of being killed by a grizzly in Yellowstone from "Bear-Inflicted Human Injuries and Fatalities in Yellowstone," National Park Service, September 18, 2019, https://www.nps.gov/yell/learn/nature/injuries.htm.

2. Werner Herzog as quoted in "Our Deep Need for Monsters That Lurk in the Dark," BBC, June 22, 2015.